Risk Management in Construction - Recent Advances
http://dx.doi.org/10.5772/intechopen.107735
Edited by Hasan Tosun, Necmi Gürsakal and Asli Sebatli-Saglam

Contributors
Osamu Takahashi, Iturou Yoshizawa, Kamoli Adetunji, Mohd Saidin Misnan, Mohamad Zahierruden Ismail, Farah Nadiah Abdul Rahim, Zulkiflee Abdul-Samad, Naimah Muhammed-Yakubu, Christopher Amoah, Christopher Sikhupelo, A.M. Faten Albtoush, Adesina Emmanuel Aladeloba, Godwin Iroroakpo Idoro, Rouzbeh Shabani, Ole Jonny Klakegg, Agnar Johansen, Olav Torp, Hasan Tosun

First published in London, United Kingdom, 2024 by IntechOpen
IntechOpen is the global imprint of INTECHOPEN LIMITED, registered in England and Wales, registration number: 11086078, 5 Princes Gate Court, London, SW7 2QJ, United Kingdom
Printed in Croatia

British Library Cataloguing-in-Publication Data
A catalogue record for this book is available from the British Library

Additional hard and PDF copies can be obtained from orders@intechopen.com

Risk Management in Construction - Recent Advances
Edited by Hasan Tosun, Necmi Gürsakal and Asli Sebatli-Saglam
p. cm.

This title is part of the Civil Engineering Book Series, Volume 2
Topic: Construction Engineering
Series Editor: Assed Haddad
Topic Editor: Amin Akhnoukh

Print ISBN 978-1-83769-309-2
Online ISBN 978-1-83769-310-8
eBook (PDF) ISBN 978-1-83769-311-5
ISSN 3029-0287

For EU product safety concerns:
IN TECH d.o.o., Prolaz Marije Krucifikse Kozulić 3, 51000 Rijeka, Croatia,
info@intechopen.com or visit our website at intechopen.com.

IntechOpen Book Series
Civil Engineering
Volume 2

Aims and Scope of the Series

Civil engineering is a traditional field of engineering from which most other branches of engineering have evolved. It comprises traditional sub-areas like transportation, structures, construction, geotechnics, water resources, and building materials. It also encompasses sustainability, risk, environment, and other concepts at its core. Historically, developments in civil engineering included traditional aspects of architecture and urban planning as well as practical applications from the construction industry. Most recently, many elements evolved from other fields of knowledge and topics like simulation, optimization, and decision science have been researched and applied to increase and evolve concepts and applications in this field. Civil engineering has evolved in the last years due to the demands of society in terms of the quality of its products, modern applications, official requirements, and cost and schedule restrictions. This series addresses real-life problems and applications of civil engineering and presents recent, cutting-edge research as well as traditional knowledge along with real-world examples of developments in the field.

Meet the Series Editor

Professor Assed N. Haddad is a Civil Engineer with a degree from the Federal University of Rio de Janeiro (UFRJ) earned in 1986, as well as a Juris Doctor degree from the Fluminense University Center earned in 1993, and a Master's degree in Civil Engineering from the Fluminense Federal University (UFF) obtained in 1992. He completed his Ph.D. in Production Engineering from COPPE / Federal University of Rio de Janeiro in 1996. Professor Haddad's academic pursuits have taken him to postdoctoral stays at the University of Florida, USA in 2006; at the Universitat Politècnica de Catalunya, Spain in 2010; and at the University of New South Wales Sydney, Australia in 2019. Currently, he serves as a Full Professor at the Federal University of Rio de Janeiro. He has held visiting professorships at various institutions including the University of Florida, Universitat Politècnica de Catalunya, Universitat Rovira i Virgili, and Western Sydney University. His research expertise encompasses Civil, Environmental, and Production Engineering, with a primary focus on the following topics: Construction Engineering and Management, Risk Management, and Life Cycle Assessment. He has been the recipient of research grants from the State of Rio de Janeiro, Brazil: CNE FAPERJ from 2019 to 2022 and from 2023 to 2025. Additionally, his research grants obtained from the Brazilian Government CNP since 2012 last to this date. Professor Haddad has been involved in several academic endeavors, being the Guest Editor of the International Journal of Construction Management; MDPI's Sustainability, Energies, and Infrastructures; Associate Editor at Frontiers in Built Environment / Sustainable Design and Construction; Guest Editor at Frontiers in Built Environment / Construction Management; and Academic Editor of the Journal of Engineering, Civil Engineering Section of Hindawi. He is currently a Professor of the Environmental Engineering Program at UFRJ and the Civil Engineering Program at UFF.

Meet the Volume Editors

Hasan Tosun is a Professor of Geotechnical Engineering. Currently, he is working at the Mudanya University, Türkiye. Dr. Tosun specializes in geotechnics for dam engineering, especially for embankment and rockfill dams, and teaches soil mechanics, geotechnics for dam engineering, and engineering geology. Up to 1997, he worked at the General Directorate of State Hydraulic Works and supervised the geotechnical studies of large dams constructed in Turkey. He has published more than 320 technical papers in national and international journals and conference proceedings and is the author of four books on soil mechanics and geotechnics for dams. He is the president of the Turkish Society on Dam Safety and a member of the National Society of Soil Mechanics and Foundation Engineering in Turkey, the United States Society on Dams (USSD), and the Canadian Dam Association (CDA).

Necmi Gürsakal is a statistics professor at Mudanya University, Türkiye, where he transfers his experience and knowledge to his students. Previously, he worked as a faculty member in the Econometrics Department, Bursa Uludag University, Türkiye, for more than 40 years. Dr. Gürsakal has published many books and articles on data science, machine learning, artificial intelligence, social network analysis, and big data. In addition, he has served as a consultant to various business organizations.

Asli Sebatli-Saglam has a Ph.D. in Industrial Engineering from Bursa Uludag University. She is currently an assistant professor at Mudanya University, Türkiye. Her master's thesis was on developing a decision support system within the scope of disaster operations management. In her Ph.D. thesis, she concentrated on developing a parking management system for smart cities in the field of intelligent transportation systems. Dr. Sebatli-Saglam's research interests include operations research applications in various areas.

Contents

Preface

In engineering, risk management is the first step to developing economical and safe solutions. Risk management is a fundamental concept in construction to realize the optimal project. In technical terms, risk is defined as the measure of the probability and severity of adverse effects on life, health, property, or the environment. Researchers consider the fact that risk is estimated by the combined impact of the scenario, probability of occurrence, and the associated consequence. In some cases, the term "risk" focuses on economic concerns, while at other times it refers to loss of life.

Risk management is the systematic process of analyzing, identifying, and responding to project risks. It mainly consists of maximizing the chances and the impact of positive events. It also helps to minimize the probability and the impact of negative events. In addition, it aims to meet the project's objectives. In other words, risk management can be considered a decision-making process, involving a full understanding of known risks and/or necessary actions to reduce the effects and chances of such risks. In the construction industry, risk management studies begin with risk identification, continue with risk analysis and evaluation, and finalize with risk response. In summary, construction risk management is the process of evaluating and implementing procedures to reduce the impact of risks in construction projects.

In this book, researchers and engineers have authored chapters focused on the latest advances in construction risk management. The book includes both theoretical and practical papers of high technical standards for various aspects of risk management, thus facilitating an awareness of modern techniques and methods. We selected high-quality submissions of original research as well as review articles that have the potential for practical application.

Chapter 1 presents a study on the risk management of large dams in the operation stage using actual projects in Türkiye. Chapter 2 presents a building damage estimation and renovation proposal system using constant microtremor measurement for construction risk management. Chapter 3 deals with recent advancements in the management of construction risks. Chapter 4 discusses managing uncertainty in the construction phase of road projects. Chapter 5 focuses on the risk factors affecting public infrastructure projects. Chapters 6 and 7 contain studies on approaches to improving occupational health and safety of the Nigerian construction industry and the risk of delay in construction projects, respectively. Finally, Chapter 8 introduces a study done on client type and communication practice during the pre-contract phase of construction project development in Nigeria.

It is our hope that this book proves a useful resource for readers studying risk management in engineering, as it introduces recent scientific research and the latest knowledge of the construction industry. We are thankful to the authors for their excellent contributions.

Dr. Hasan Tosun and Dr. Asli Sebatli-Saglam
Faculty of Engineering, Architecture and Design,
Department of Industrial Engineering,
Mudanya University,
Bursa, Türkiye

Dr. Necmi Gürsakal
Faculty of Arts and Social Sciences,
Department of Psychology,
Mudanya University,
Bursa, Türkiye

Chapter 1

Risk Management of Large Dams under Operation Stage and Case Studies for Cascade System

Hasan Tosun

Abstract

In recent specifications, there are regulations stating that there should be no accidents on construction sites. In fact, construction-related risks have cultural, social, and technological dimensions, and risks can be minimized by appropriate identification of these factors. Although there are no great social and cultural uncertainties in the applications carried out so far, the technological dimension, especially the use of digital technologies, has been at a minimum level. Currently, the construction industry investigates how social-technical systems interact during the construction of all large engineering structures, including dam structures, and their effects on risks at construction sites. However, after the completion of the construction of these structures, management of the risk posed by the relevant structures for the environment is a very separate issue. This study focuses on the total risk of dam structures that have been constructed and are in the operational stage. Due to damage or failure of a dam, downstream life is affected and huge hazards may occur. In fact, major disasters may occur depending on the amount of water to be discharged. For this purpose, the risk of dams is defined according to their type, height, amount of water they store, age of the dam, downstream evacuation requirement, and potential damage area. In newly presented methods, local ground effects are also included. In addition, the fact that the relevant structure is in a cascade system affects the degree of hazard that will occur and increases the total risk of the system. In this study, the parameters affecting the total risk of dam structures in the operational stage are summarized, various total risk methods are mentioned, the total risks for two separate rivers located in the largest basin of the country are discussed and the necessary measures to be taken on the basis of risk management are suggested. Finally, a new concept is introduced for risk management of dams in the cascade system.

Keywords: risk management, dam safety, total risk, seismic hazard, river basin

1. Introduction

Every construction project, regardless of size, scope, complexity, nature, or location, has unique risks. Failure to manage relevant projects correctly also increases the risk and causes losses to the balance sheet. It should be emphasized that a risk, even if small, in the project may create uncertainties during the implementation phase and lead to negative consequences that may affect the success of the project. Additionally, new risks may arise at different stages of a project. However, it is clear that high risks waste resources, increase project costs, reduce efficiency, and extend project completion time. Therefore, it is of great importance to correctly identify and manage risks at every stage of the project. Risk definitions are of greater importance in the management of water resource structures such as dams and storage facilities, which are built for public service purposes and are vital for living life downstream.

Although dams provide benefits such as domestic and industrial water supply, electrical energy production, and agricultural irrigation, they may cause some environmental impacts. These affect communities through resettlement and other socio-economic impacts, result in environmental concerns, and create sedimentation problems in the long term. In addition to all the negative consequences, if they collapse, they can cause major disasters resulting in great loss of life and property. These problems and concerns need to be alleviated through proper planning and incorporating various improvement techniques according to the demands of society. Hariri-Ardebeli [1] suggests that these expectations can be most effectively met by implementing systematic risk management that includes factors such as sustainability and public participation during dam planning, construction, and operation stages.

According to 2016 data in the USA, there are over 90,000 small or large dams and approximately 18% of them are defined as high-hazard dams. It has also been observed that the average lifespan of dams is over 50 years. It has been stated that at least 64 billion dollars are needed to meet the current safety requirements of these dams across the country [1]. The number of large dams in Turkey is close to 1500 and their average age exceeds 30 years. These dams have costs of up to 10 billion dollars for changing physical conditions and updated safety requirements. Significant damage occurred in 40 dams by two major earthquakes (Mw7.8 and Mw7.6) on February 06, 2023 [2, 3] and the cost of the damage to these dams was determined as 4.0 billion dollars [4]. All these developments necessitate the effective use of water resources for social life (water and food safety as well as the danger of downstream life), and the need for all risks (especially during the operation phase) to be well identified and managed effectively.

Safety concern for a dam structure is mainly affected by ground motion and result in loss of stability and some physical defects in the structure [5]. Pre-planning for dams and earthquake mitigation generally uses simplified procedures. An assessment of the overall stability of the structure should be included. Simplified procedures are generally used for pre-planning earth-rockfill dams. The analysis of concrete dams exposed to earthquakes must include an assessment of the stability of the entire structure to withstand lateral forces and moments [6, 7]. Various methods from simplified analysis to sophisticated techniques are used to analyze seismic loads acting on concrete structures. It should be noted that seismic analysis is the first step in determining the overall risk of a dam structure.

This chapter considers a portfolio of dams on two tributaries of the main river of Euphrates Basin. The main River, with its tributaries of Murat, Perisuyu, and Karasu has a 2800 km length. It flows through Iraq into the Gulf in the south. The flows upstream and on the Syrian border are 650 cubic meters per second and 950 cubic meters per second, respectively. Within the basin, 40 large dams are constructed to use the basin's energy and irrigation potential. This article assesses the overall risk of 11 large dams in the two Euphrates River cascade systems. In other words, this study summarizes the safety of dam structures during the operational phase, especially

#	Dam	River	Height from river bed (m)	Function (*)	Type (**)	Completed year	Reservoir capacity (hm^3)	Installed capacity (MW)
1	Keban	Euraphate	163	E + F + I	CG + RF	1975	31,000.0	1330
2	Agri-Yazici	Murat	84	I + WS	EF	2001	202.9	—
3	Alparslan-1	Murat	91	E	RF	2009	2993.0	160
4	Alparslan-2	Murat	108	E + I + WS	ACEF	2020	2431.0	280
5	A.Kalekoy	Murat	93	E	CG + ACRF	2020	516.5	500
6	Beyhan-1	Murat	84	E	RCC	2015	369.1	582
7	Beyhan-2	Murat	42	E	CG	—	78.9	264
8	Gayt	Murat	36	I	EF	1998	23.0	—
9	Gulbahar	Murat	66	I	EF	2003	14.1	—
10	Patnos	Murat	38	I	EF	1992	33.4	—
11	Y.Kaleköy	Murat	138	E	RCC	2018	783.8	636.6
12	Kalecik	Peri	34	I	EF	1987	12.5	—
13	Kigi	Peri	146	E	RF	2016	507.6	140
14	Konaktepe	Peri	119	E	RF	u/d***	450.0	201
15	Ozluce	Peri	124	E	RF	1998	1075.0	170
16	Pembelik	Peri	75	E	HF	2015	358.0	127
17	Seyrantepe	Peri	34	E	EF	2008	23.0	56
18	Tatar	Peri	73	E	EF	2013	300.0	128
19	Uzuncayir	Peri	58	E	EF	2009	308.0	82

***RF: Rockfill EF: Earthfill CG: Concrete gravity HF: Hardfill ACRF: Rockfill with asphaltic core ACEF: Earthfill with asphaltic core.*
*****Under designing.*

Table 1.
Characteristics of large dams on two tributaries of Euphrates [8].

with regard to the risks to life downstream, highlights the methods used for the overall risks of the structures involved, and highlights the risks posed by dams.

The author discusses individual structures and cascading structures in the country's largest watershed. **Table 1** introduces the general characteristics of dams. This paper refers to some studies done previously [9–11]. The author and his co-workers have completed many international and national projects about dam safety and risk management of earth structures relating to water resources and introduced a lot of studies in congress, symposium, and journals. These studies mainly concentrate on dam safety concepts in majoring and risk management of dams and their appurtenant structures in minoring [12–28].

2. Methods and materials

Risk management is a key issue in dam construction [1]. Technically speaking, risk is described as the likelihood and severity of adverse impacts acting on life, environment, and property. According to an international specification [29], the risk is generally predicted by combining the probability of its occurrence and the impact of the scenario on the associated outcome and defined by a simple formula (Eq. (1)). In this equation, P means conditional probability depending on variables X and Y and also C represents outcomes [9].

$$\text{Risk} = \int P\left[\text{Load Events}\right] \times P\left[\text{Responses} \mid \text{Loads}\right] \times C\left[\text{Loads, Responses}\right] \qquad (1)$$

In the construction industry, the term "risk" can primarily refer to economic concerns, but can also refer to life loss. Technical specifications mostly use a "risk curve" to measure safety. Seismic hazard, on the other hand, represents the uncertain relationship between seismic intensity level and the probability that at least excitation level will occur at the given location. The hazard curve represents a graph that correlates the seismic intensity of a particular location on the horizontal axis and the annual exceedance frequency in vertical axis [7].

There are various techniques to quantify a dam's overall risk factors. The ICOLD method takes into account seismic hazards belonging to dam site and structure's risk separately [30]. In this method, the seismic hazard can be classified into four groups, regardless of the dam type [31, 32]. This is an easy manner to evaluate earthquake hazard via way of means of figuring out the hazard class. ICOLD introduces that the whole risk of dams includes structural and socio-monetary factors [33]. The first is especially based upon at dam height and ability of the reservoir. The second one is the capability for downstream damage and evacuation requirements. The summation of these four risks represents the Risk Factor (RF). Four danger classes, which can be used for deciding on seismic parameters of floor motion, are described on the idea of the RF values. ICOLD states that big dams (peak is extra than 90 meters and a storage potential of extra than 1200 hm^3) require unique protection considerations [30].

In general, a dam hazard evaluation must be decided primarily based totally on reservoir capacity, dam height, evacuation requirements, and capability downstream damage. Generally, seismic evaluation and risk evaluation are evaluated separately. However, Bureau [33] combines those elements to outline the overall hazard factor for dam structures (TRF), as given in Eq. (2). In the equation, TRF refers to total risk factor. The first three phrases cowl dam-based elements, and HRF covers dam-based elements. Downstream risk as a feature of population and properties at risk is described as a feature of specific elements (Eq. (3)) [34, 35]. In this equation, ERF is the evacuation demand component and DRI is the downstream damage risk index. The Downstream Evacuation Requirement (ERF) component relies upon the population at risk. The Downstream Damage Risk Index is primarily based on the value of private, commercial, and industrial properties in the capability flood path. Finally, the PDF shows the anticipated damage component (Predicted Damage Factor), which is a feature of the dam's determined overall performance in addition to the location-based seismic hazard.

$$\mathrm{TRF} = \left[(\mathrm{CRF} + \mathrm{HRF} + \mathrm{ARF}) + \mathrm{DHF}\right] \times \mathrm{PDF} \tag{2}$$

$$\mathrm{DHF} = \mathrm{ERF} + \mathrm{DRI} \tag{3}$$

Vulnerability evaluation represents the location-based seismic risk described by the Predicted Damage Factor (PDF), which is assigned to every dam consistent with Eq. (4). In this equation, PDI is the Predicted Damage Index, calculated the usage of the dam fragility curve advanced with the aid of using Bureau and Ballentine [36]. The PDI relies upon the dam type and the seismic risk of the dam site and is calculated as a feature of the Earthquake Severity Index (ESI), which represents the predicted ground motion on the dam site as given with the aid of using Eq. (5). In this equation PGA is peak ground acceleration in g; and M is the Richter or moment magnitude.

$$\mathrm{PDF} = 2.5 \times \mathrm{PDI} \tag{4}$$

$$\mathrm{ESI} = \mathrm{PGA} \times (\mathrm{M} - 4.5) \tag{5}$$

The author proposed a qualitative evaluation framework that shows the relative impact of dams within a cascade system. Five different risk rates (very low, low, moderate, high, and very high) are defined for every structure in the system. The principles are listed in **Table 2** according to dam type, reservoir storage capacity, and dam location. All dams fall into two basic groups: "rigid" and "flexible." "Flexible" refers to earth-fill dams, rock-fill dams, and combinations thereof; "rigid" refers

#	Dam type*	Number of dams that are potentially under failure downstream (N)	Risk ratio for cascade system (RRCS)
1	Rigid	—	Very low
2	Flexible	—	Low
3	Flexible Rigid	<2 1–3	Moderate
4	Flexible Rigid	2–3 3–4	High
5	Flexible Rigid	≥3 ≥4	Very high

**Flexible means earthfill dams, rockfill dams, and their combination while rigid describes concrete dams (gravity, ach, buttress), roller compacted dams and hardfill dams.*

Table 2.
Risk ratio for dams of a cascade system.

to concrete dams (gravity, arch, and buttress), roller-compacted concrete dams, and hardfill dams. Another factor considered is the number of downstream dams that may fail. This is based on the relative storage capacity of the dam. The dam risk assessment considered how much damage downstream dams would suffer if the dam completely collapsed. This model is based on the number of dams in the downstream area and assumes that the rigid dam acts as an overall spillway structure and is not damaged. For composite types, it may make sense to consider weaker ones. The author first mentioned this problem in ICOLD symposiums held in Prague and Chania in 2017 and 2019, respectively [37, 38] and in a recent study published in an international book project [39]. This problem was also considered by some scientists from east-south Asia [40, 41].

3. Case studies and discussion

In the study, risks in dams that are in operation are discussed and risk management of dams in individual and cascade systems is emphasized. For this purpose, some dams on the Upper Euphrates river located in the east of Turkey were examined by considering two separate cascade systems. There are 11 dams on these two separate systems, located on the Murat river and Perisuyu tributaries in the Upper Euphrates. The country's second-largest reservoir (Keban dam), and two dams on the Munzur stream, which is considered a part of the Perisuyu system have also been included to the system considered in the study. Individual and cascade systems of dams were evaluated on the basis of total risks and recommendations were made for risk management. All analyses were carried out depending on the seismo-tectonic models in the regions [42–44]. The projects and their results, which were done previously, have been considered in this study to describe the problem in this study [20, 45, 46].

The first cascade system is located on the Murat river. This system records a total of six separate large dams having heights between 42 and 127 m. The structures located on the main river from downstream to upstream are Beyhan-2, Beyhan-1, Lower Kalekoy, Upper Kaleköy, Alparslan-2, and Alparaslan-2 dams (**Figure 1a**). The total installed power of these dams, which were built mainly for energy production, is 2337 MW and it is planned to produce 5740 GWh of electrical energy annually. In addition, together with some other dam structures on the tributaries, it is envisaged to irrigate an area reaching 120,000 ha. Two large dams located upstream were constructed in embankment type (rock fill and asphaltic core rockfill) and the other four dams were in rigid type (concrete gravity, roller compacted dam). **Figure 1a** shows locations and some basic features of the cascade structure of the dams. The related data about dam characteristics are given in **Table 1**.

The system evaluated in this study and forming the second cascade structure of the basin is located on Perisuyu. These structures, whose heights vary between 34 and 146 m; It consists of Kığı, Özlüce, Pembelik, Seyrantepe, and Tatar dams from upstream to downstream. Two separate dams (Konaktepe and Uzunçayır) on the Munzur stream, which constitutes an important side branch of this system, were also considered as a part of the system. All of these dam structures, except one (Pembelik dam), were planned as fill type (rock fill and earth fill). The total installed power of these dams is 624 MW and they have the potential to produce 1766 GWh of electrical energy annually (**Table 1** and **Figure 1b**).

First of all, seismic hazard analyses were carried out for all dams considered in the study and the seismic parameters that should be taken into account in terms

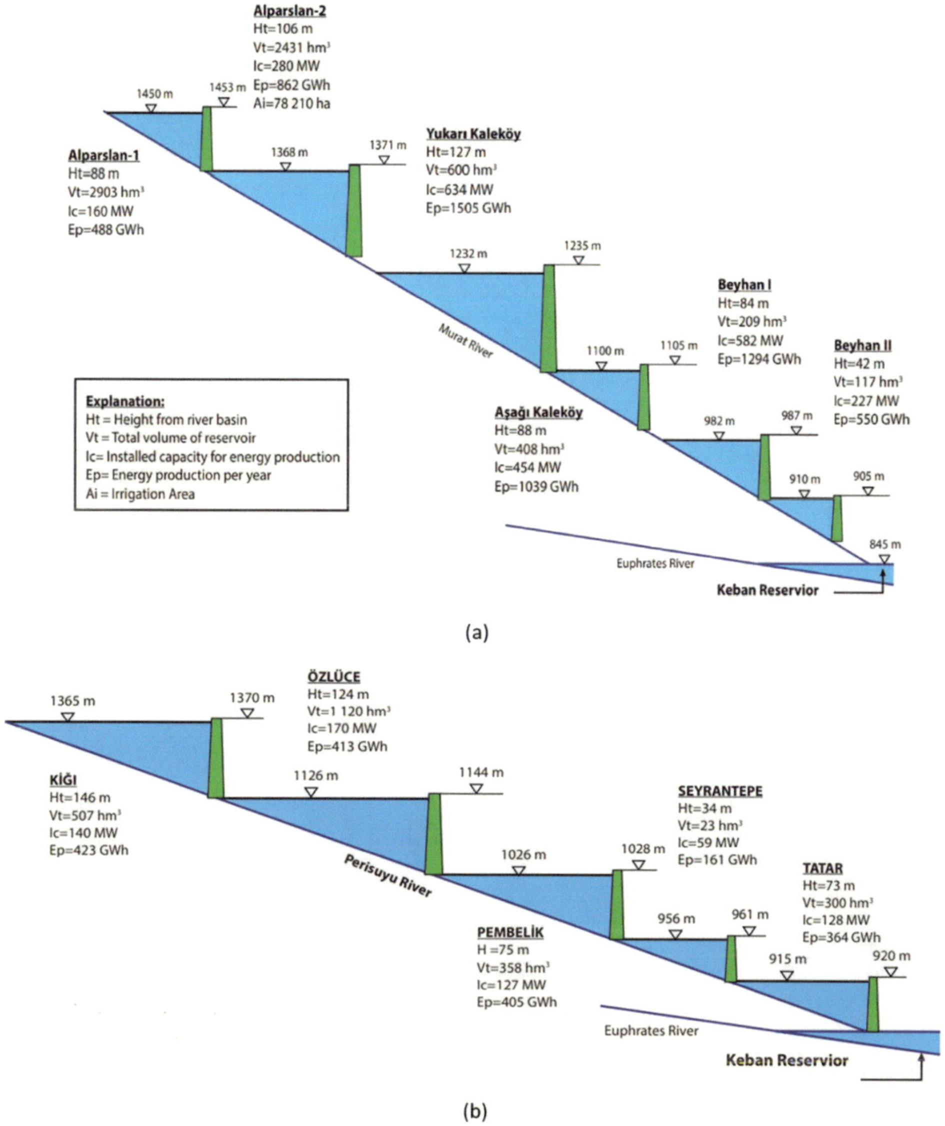

Figure 1.
Cascade systems of tributaries of main river of Euphrates basin: (a) Murat river and (b) Peri-Munzur river [8].

of total risk were determined. On this basis, the hazard rates for the 11 dams taken into consideration are in the “high-very high” range (**Table 3**). Based on the ICOLD method, the risk factor was 36, which is the highest value of the related classificaiton (excluding one dam), and the risk class was determined as between III and IV, and the risk rate was predicted as “high-very high.” According to the Bureau method, total risk factor values vary between 143.3 and 229.1, and the Risk Class and Risk Ratio values for all dams are III and “High,” respectively (**Table 3**).

The author believes that the risk of dams during the operation stage should be evaluated together with all elements within the cascade system. For this purpose, he introduced a new proposal. The results of this study showed that individual risk definitions of all dams within the two selected cascade systems were formed similarly.

#	Dam	Hazard analysis		Total risk [30]			Total risk [33]			Risk for cascade system*			
		Class	Hazard ratio	Risk factor	Risk class	Risk ratio	Total risk factor	Risk class	Risk ratio	V (hm^3)	N	Type	RRCS
Murat tributary													
1	Alparslan-1	IV	Very high	36	IV	Very high	148.2	III	High	2993	2	RF	High
2	Alparslau-2	III	High	36	IV	Very high	201.0	III	High	2431	1	ACRF	High
3	Y.Kaleköy	IV	Very high	36	IV	Very high	168.7	III	High	784	1	RCC	Moderate
4	A.Kalekoy	III	High	36	IV	Very high	143.3	III	High	517	—	ARCF	Low
5	Beyhan-1	IV	Very high	36	IV	Very high	173.5	III	High	369	—	RCC	Very low
6	Beyhan-2	IV	Very high	36	IV	Very high	173.5	III	High	79	—	CG	Very low
	Keban	II	Moderate	36	IV	Very high	126.4	III	High	31,000			
Munzur tributary													
1	Konaktepe	III	High	36	IV	Very high	168.0	III	High	450	1	RF	Moderate
2	Uzuncayir	III	High	36	IV	Very high	212.3	III	High	308	—	RF	Low
	Keban	II	Moderate	36	IV	Very high	126.4	III	High	31,000			
Perisuyu tributary													
1	Kigi	III	High	36	IV	Very high	176.1	III	High	508	3	RF	Very high
2	Ozluce	IV	Very high	36	IV	Very high	174.5	III	High	1075	2	RF	High
3	Pembelik	IV	Very high	36	IV	Very high	229.1	III	High	358	2	HF	Moderate
4	Seyrantepe	IV	Very high	24	III	High	162.9	III	High	23	1	EF	Moderate
5	Tatar	III	High	36	IV	Very igh	147.5	III	High	300	—	EF	Low
	Keban	II	Moderate	36	IV	Very high	126.4	III	High	31,000			

(*) V=Storage capacity at Maximum Water Level; N: Number of dams at which are potentially under failure in downstream; ype: RF = Rockfill; EF = Earthfill; CG = Concrete gravity; HF=Hardfill; ACRF = Rockfill with asphaltic core; RCC = Rollered Compacted Concrete RRCS = Risk Ratio for Cascade System.

Table 3.
Risk for individual and cascade systems for dams of the basin.

However, it is seen that the dams in the cascade system do not affect the system at the same impact value in terms of the stability of the system. Within the principles suggested by the author, among six large dams in the cascade system of Murat river, Alparslan-1 dam has and Alparslan-2 have a "high" risk rate while Yukarı Kaleköy dam poses "moderate" risk. Lower Kaleköy, Beyhan-1, and Beyhan-2 dams are represented with "low" and "very low" risk rates.

In the Perisuyu-Munzur cascade system, the risk rates of Kığı and Özlüce dams were found to be "very high" and "high," respectively. Konaktepe, Pembelik, and Seyrantepe dams have a "medium" risk rate, and Uzunçayır and Tatar dams have a "low" risk rate.

Keban dam, which is located downstream of these two cascade systems and has a large storage capacity (31,000 hm^3), has the location and capacity to compensate for the flooding resulting from all dams upstream. It is necessary to realize a specific program of operation for Keban dam and others located downstream to provide a stable condition for all system upstream.

4. Conclusions

Risk management during the operation phase of dam structures differs from the risk management of other engineering structures. Therefore, instead of independent risk assessments of structures, it is appropriate to evaluate the risk of the cascade structure as a whole. The safety of cascade dams must be considered with a new concept from that of single dams. Collapse of a cascaded dam can lead to flooding or collapse of downstream dams. The author assumes that the probability of dam failure within a cascade structure is greater than that of a single dam. Concrete dams cannot fail when they are overtopped by flood waves. Because they operate as a spillway during flooding. There are so many river power plants constructed as concrete structures, which had a functional operation, without damages or minor damages after the Wenchuan earthquake in 2008. Similar cases have been observed in Kahramanmaraş earthquakes with Ms. of 7.6 and 7.8 occurred on February 06, 2023. An embankment dam can fail to release reservoir water when it is overtopped by a flood wave. Therefore, embankment dams in cascade systems are more dangerous structures when compared with concrete dams for unexpected cases. Within the scope of this study, a risk identification has been made for the cascade structure and a qualitative assessment basis that reveals the most critical structure(s) in the system has been proposed. For the study, Özluce and Kigi dams are the critical structures when considering the Perisuyu-Munzur cascade system while Alparslan-1, Alparslan-2, and Y. Kaleköy dams are ones for Murat River cascade system of Upper Euphrates basin. The risk concept for the cascade system of dams, which was proposed in this study should be developed in the future. The author states that the proposal is just a beginning stage to start the discussion on risk management for dams of cascade systems in dam engineering.

Conflict of interest

The author declares that there is no conflict of interest.

Author details

Hasan Tosun
Mudanya University, Bursa, Türkiye

*Address all correspondence to: hasan.tosun@mudanya.edu.tr

References

[1] Hariri-Ardebeli MA. Risk, reliability, resilience (R3) and beyond in dam engineering. International Journal of Disaster Risk Reduction. 2018;**31**:806-831. DOI: 10.1016/j.ijdrr.2018.07.024

[2] Tosun H. Preliminary assessment report on dam damages resulted by Pazarcik (Mw 7.7) and Elbistan (Mw7.6) earthquakes of Türkiye, occurred on February 06, 2003 and potential problems in the future. 2023. DOI: 10.13140/RG.2.2.21644.80001

[3] Tosun H. Second Report-Pazarcik (Mw 7.7) and Elbistan (Mw7.6) earthquakes dated February 06, 2023: Evaluation on damages inspected on dams in the region. 2023. DOI: 10.13140/RG.2.2.19846.65601

[4] CBS. 2023 Kahramanmaraş and Hatay Earthquakes Report. Ankara: Presidency of the Republic of Türkiye; 2023. 143 p

[5] ICOLD. Selecting Seismic Parameters for Large Dams-Guidelines. International Commission on Large Dams (ICOLD); Bulletin 148, 2016

[6] Hariri-Ardebili MA, Saouma V. Probabilistic seismic demand model and optimal intensity measure for concrete dams. Structural Safety. 2016;**59**:67-85

[7] Hariri-Ardebili MA, Nuss LK. Seismic risk prioritization of a large portfolio of dams. Revisited. Advances in Mechanical Engineering. 2018;**10**(9):1-20. DOI: 10.1687814018802531

[8] DSI. Dams of Turkey. In: The General Directorate of State Hydraulic Works. Ankara: DSI; 2014. 588 p

[9] Tosun H, Tosun VT, Hariri-Ardebili MA. Total risk and seismic hazard analysis of large embankment dams: Case study of northwest Anatolia, Turkey. Life Cycle Reliability and Safety Engineering. 2020;**9**:329-338. DOI: 10.1007/s41872-020-00113-4

[10] Tosun H, Zorluer İ, Orhan A, Seyrek E, Savaş H, Türköz M. Seismic hazard and total risk analyses for large dams in Euphrates basin, Turkey. Engineering Geology. 2007;**89**(1-2):155-170

[11] Tosun H, Türkoz M, Savas H. River basin risk analysis. International Water Power & Dam Construction. 2007;**V**:30-37

[12] Tosun H. Earthquake-Resistant Design for Embankment Dams. Ankara, Turkey: Publication of General Directorate of State Hydraulic Works; 2002 (in Turkish)

[13] Tosun H. Total risk analysis of dam and appurtenant structures in a basin and a case study. In: International Congress in River Basin Management. Vol. I; 22-24 March; Antalya. 2007. pp. 477-488

[14] Tosun H. Evaluating earthquake safety for large dams in southeast Turkey. Hydro Review Worldwide (HRW). 2008;**16**(4):34-40

[15] Tosun H, Savas H. Seismic hazard analyses of concrete dams in Turkey. In: Proceeding of CDA Conference. 2005. pp. 1-7

[16] Tosun H, Turkoz M. Total risk-analyzing methods for dam structures and a case study in Turkey. In: Proceeding of CDA Annual Conference. 2007. pp. 1-9. Paper No. 1638

[17] Tosun H, Seyrek E. Total risk analyses for large dams in Kizilirmak

basin, Turkey. Natural Hazards and Earth System Sciences. 2010;**10**(5):979

[18] Seyrek E, Tosun H. Deterministic approach to the seismic hazard of dam sites in Kızılırmak basin, Turkey. Natural Hazards. 2011;**59**(2):787

[19] Seyrek E, Tosun H. Influence of analysis methods for seismic hazard on total risk of large concrete dams in Turkey. Journal of the Faculty of Engineering and Architecture of Gazi University. 2013;**28-1**:67-75

[20] Tosun H. Earthquakes and dams. In: Moustafa A, editor. Earthquake Engineering-From Engineering Seismology to Optimal Seismic Design of Engineering Structures. London, UK: IntechOpen; 2015. DOI: 10.5772/59372

[21] Tosun H. Safety assessment of large reservoir constructed for domestic water near urban areas and a case study. In: Proceeding of ICOLD-ATCOLD Symposium on Hydro Engineering. Wien: ICOLD-EUROCLUB; 2018. pp. 917-927

[22] Tosun H. Seismic stability of large dams located near energy source and a case study. In: Proceeding of 5th International Conference on Earthquake Engineering and Seismology; 8-11 October 2019; METU-Ankara

[23] Tosun H. Earthquake safety evaluation for large dams located near the energy source and case studies. In: Proceeding of 11th ICOLD European Club Symposium; 2-4 October 2019; Chania, Crete

[24] Tosun H. Hazard and total risk analyses of large dams under threat of the north Anatolian Fault Zone in mid-Anatolia, Turkey. In: Proceeding of 5th World Congress on Civil, Structural, and Environmental Engineering (CSEE'20) Virtual Conference; October 2020. DOI: 10.11159/icgre20.191

[25] Tosun H. Experience on earthquake safety of large embankment dams constructed in Turkey. In: Proceedings of the 6rd International Conference on Civil Structural and Transportation Engineering (ICCSTE'21), Virtual Conference; 17-19 May 2021. DOI: 10.11159/iccste21.137

[26] Tosun H. Total risk and seismic stability of existing large CFRD's in Turkey. In: Proceedings of the 6rd International Conference on Civil Structural and Transportation Engineering (ICCSTE'21) Virtual Conference; 17-19 May 2021. DOI: 10.11159/iccste21.138

[27] Tosun H. Dynamic stability of large embankment dams and a case study. In: Proceedings of the 6rd International Conference on Civil Structural and Transportation Engineering (ICCSTE'21) Virtual Conference; 17-19 May 2021. DOI: 10.11159/iccste21.139

[28] Tosun H. Safety evaluation for older large dams in Turkey. In: ICOLD Symposium on Sustainable Development of Dams and River Basins; 4-27 February 2021; New Delhi. 2021

[29] ICOLD. Bulletin on Risk Assessment in Dam Safety Management. Technical Report. Paris, France: International Commission on Large Dams; Bulletin 130, 2005

[30] ICOLD. Selecting Parameters for Large Dams-Guidelines and Recommendations. In: ICOLD Committee on Seismic Aspects of Large Dams, Bulletin 72. (2010 Revision), 1989. p. 37

[31] Tosun H. Earthquake safety assessment of dam structures. In: Lecture Notes of Dam Safety Workshop by Water

Foundation (Werdec); İstanbul. 2007. pp. 95-106

[32] Tosun H. Seismic hazard analyses and a seismo-tectonic model for Turkey. In: 5th Congress on Dams; September 2021; Struga, Republic of N. Macedonia. pp. 1-6

[33] Bureau GJ. Dams and appurtenant facilities. In: Chenh WF, Scawthorn C, editors. Earthquake Engineering Handbook. Bora Raton: CRS Press; 2003. p. 26. 1-26. 47

[34] Tosun H. Dam safety classification and their related criteria. In: Lecture Notes of Dam Safety Workshop by Water Foundation (Werdec); İstanbul. 2007. pp. 107-115

[35] Topçu S, Tosun H. An overview on total risk classifications for dams. In: 5th Congress on Dams; September 2021; Struga, Republic of N. Macedonia. pp. 1-9

[36] Bureau G, Ballentine G. A comprehensive seismic vulnerability and loss assessment of the state of south Carolina using Hazus. Part VI. Dam inventory and vulnerability assessment methodology. In: 7th US National Conference on Earthquake Engineering; Boston, Massachusetts, USA. 2002. pp. 1943-1953

[37] Tosun H, Oguz S. Stability analysis of Atatürk dam, Turkey as based on the updated seismic data and design code. In: Proceeding of 85th Annual Meeting of International Commission on Large Dams, Prague. 2017. pp. 66-68

[38] Tosun H, Oguz S. Seismic design for existing large dams and case studies. In: Proceeding of 11th ICOLD European Club Symposium; 2-4 October 2019; Chania, Crete

[39] Tosun H. Recent evaluation on total risk of cascade dams on Murat River of Upper Euphrates Basin, Turkey. In: Special Topics in Dam Engineering. London, UK: IntechOpen; 2022. Available from: https://www.intechopen.com/chapters/1081439

[40] Chen Y, Lin P. The total risk analysis of large dams under flood hazards. Water. 2018;**10**:140. DOI: 10.3390/w10020140

[41] Zhang R, Wang B. Seismic hazard and total risk analysis for cascade dams situated in geologically sensitive regions. Research Square. 2022. DOI: 10.21203/rs.3.rs-1985090/v1

[42] Fraser WA, Howard JK. Guidelines for Use of the Consequence-Hazard Matrix and Selection of Ground Motion Parameters. California: Technical Publication, Department of Water Resources, Division of Safety of Dams; 2002

[43] Jiminez MJ, Giardini D, Grünthal G. Unified seismic hazard modelling throughout the Mediterranean region. Bolettino di Geofisica Teorica ed Applicata. 2001;**42**(1-2):3-18

[44] Tosun H. Stability Analysis of Large Embankment Dams Located on Shear Zones and Case Studies. Denver: Hydrovision International, Colorado Convention Center; 2017

[45] Tosun H. Re-analysis of Ataturk dam under ground shaking by finite element models. In: Proceeding of CDA Annual Conference; 22-27 September; Saskatoon, Canada. 2011. pp. 1-11

[46] Tosun H. Earthquake safety of Keban dam, Turkey. In: Proceeding of CDA Annual Conference; 15-20 October; Fredericton, NB, Canada. 2012. pp. 1-12

Chapter 2

Perspective Chapter: Building Damage Estimation and Renovation Proposal System Using Constant Microtremor Measurement for Construction Risk Management

Itsuro Yoshizawa and Osamu Takahashi

Abstract

In this chapter, the authors propose a system that measures microtremors of the ground and buildings and instantaneously calculates building damage estimation and emergency risk assessment. In the conventional damage estimation and emergency risk assessment due to microtremors, resonance judgments are mainly based on the initial state due to microtremors of the ground, and buildings are often seen. However, there is a possibility that the ground and the building will resonate continuously as the ground and the building are cracked and damaged by disturbances such as earthquakes and winds. In addition, the resonance causes great damage to the building. In the system proposed this time, damage estimation and emergency risk assessment are performed not only in the initial state but also during the duration. In addition, as a renovation proposal, a system is proposed that predicts the damage and emergency risk of the building during renovation with earthquake resistance, seismic isolation, and damping. In this system, the building hazard levels are classified into four categories based on two methods: one method involves calculating the response spectrum and transfer function of the building and ground using continuous microtremor measurements to determine the resonance performance, and the other method utilizes the PML (Probable Maximum Loss) values.

Keywords: health monitoring, application, an emergency safety evaluation, constant tremor, probable maximum loss

1. Introduction

In Japan, after an earthquake, local government officials and building experts conduct building damage surveys for the purpose of preventing secondary damage and confirming the availability afterward. According to technical guidelines on the damaged building damage classification criteria and restoration, the restoration procedure is divided into three stages: (1) "This building is usable" (2) "Caution is required when entering" and (3) "Entering this building is dangerous" [1], and investigations and restoration measures according to those stages are examples of restoration strategies. It is listed as. Looking at the implementation status of past earthquakes, it takes about two weeks to complete the emergency risk assessment and about three months to complete the disaster classification assessment. The current situation is that recovery activities are further delayed due to confusion after the earthquake, shortage of people, and insufficient number of contractors. We searched for a system construction method for emergency risk determination and verified its validity. Technical Guidelines have been produced for seismic damage classification criteria on affected buildings and restoration after the disaster [1].

Earthquake preparedness guidelines and reality in Japan, the information required for building analysis based on assumed earthquakes is often accessible to the public through the Internet. The information required for building analysis based on assumed earthquakes includes, for example, the publication of borehole survey results for the specific area on websites [2]. These results can be used for earthquake analysis. Additionally, the "J-SHIS Earthquake Hazard Station" (https://www.j-shis.bosai.go.jp/: National Research Institute for Earth Science and Disaster Resilience) [3] provides a probabilistic seismic motion prediction map, which uses a finely classified terrain model in 250m mesh, to present the probability of future earthquake motion based on the prediction of seismic source faults. Data collection based on location can be retrieved from these server spaces. Furthermore, some of this information is also available as web-based APIs, indicating a trend toward systematizing their construction.

The purpose of this study is to reduce human or building damage in the event of an earthquake, clarify the dangers of individual buildings before taking concrete disaster prevention measures, and effectively call for countermeasures against earthquake damage.

Research on emergency risk assessment is often measured with a dedicated accelerometer that is not widely used, but in this research, the accelerometer built into a mobile device such as a smartphone is used, and it is generally used via Internet access [4]. Will build a system that can easily perform predictive diagnosis of emergency risk judgment. In addition, the system to be constructed will be constructed, as shown in **Figure 1**, by unifying the load on the mobile terminal, the storage of research information, and the judgment processing. We will bring the analysis method by constant

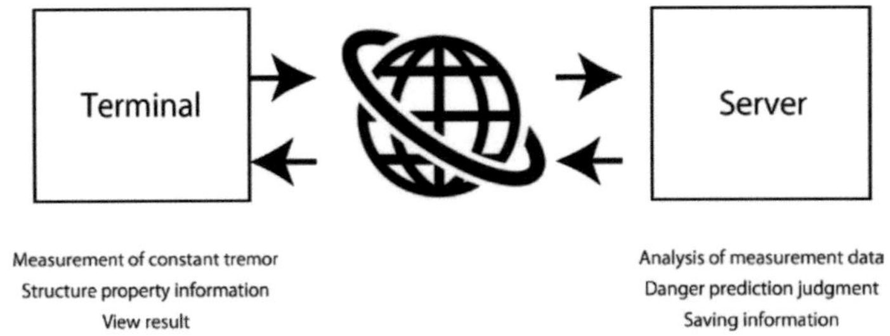

Figure 1.
Overview of damage prediction judgment system using mobile terminals.

tremor and PML to this system, actually distribute it, and consider the tendency of the judgment result about the obtained result.

2. Construction of mobile terminal system

2.1 System overview

Currently, Android, iOS, etc., are the mainstream operating systems for smartphones, and the system construction method is different for each. In this research, we focus on Android as an example (**Figure 2**), but if a similar API is used, it will be possible to develop the same for iOS.

In addition to using Android Studio in Android development, functions such as map function and geocoding use google API, and it is necessary to use google developers console to use these(**Table 1**).

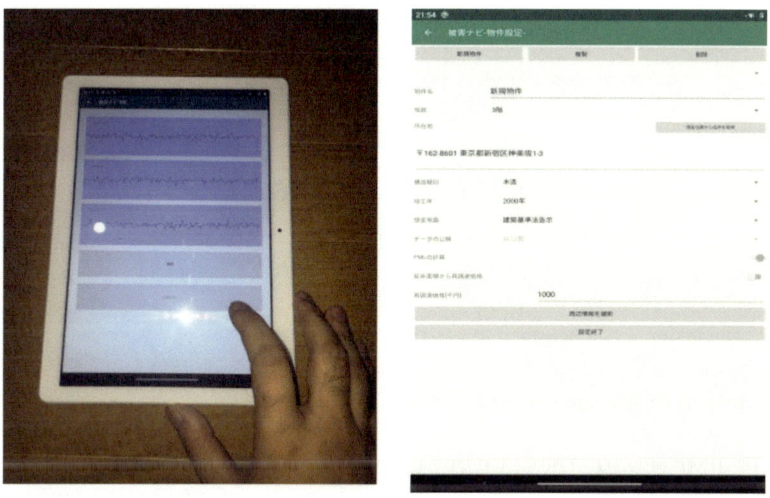

Figure 2.
Appearance of damage prediction judgment system using mobile terminals.

Function, environment	Explanation
Android	Operation system
Android studio	Platform
Google MAP	Screen using Googlemap
Firebase	Communication between server and terminal

Table 1.
Terminal development environment.

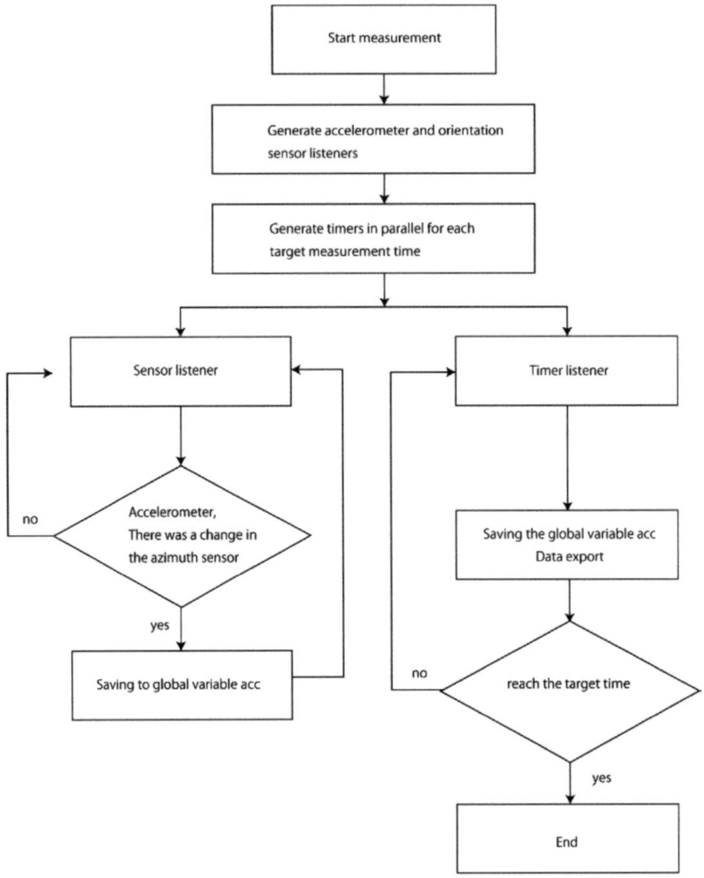

Figure 3.
Appearance of damage prediction judgment system using mobile terminals.

2.2 Construction of measurement system

It is difficult to acquire the values of the acceleration sensor and the azimuth sensor of the mobile terminal at equal intervals because both of them are postprocessed when the values are changed. Therefore, it is necessary to store it in a global variable or static variable in the posted process, and it is also necessary to post a message independently by using a timer function, etc., in parallel processing. In addition, among the processes posted by the sensor, the acceleration in the world coordinate system is acquired by performing the attitude control process of the mobile terminal. For attitude control, after acquiring the rotation matrix with the geomagnetic sensor, the inverse matrix is multiplied by the acceleration matrix to convert it into the world coordinate system. At this time, gravity is removed. The post from the timer mainly runs the export process. The flow chart of the measurement process is shown in **Figure 3**.

2.3 Creating a summary file on a mobile terminal and communicating with the server

In the terminal, it is necessary to create comprehensive data including measurement data and property information for measurement data. After transferring the file by FTP communication, the terminal creates a queue for the server to start analysis and sends it. For message queues, information is sent to SQL and processed to replace the queue.

3. Building a system for a server

3.1 System overview

The server constantly evaluates the resonance by tremors and analyzes by PML (Probable Maximum Loss) value, and makes a prediction judgment of the damage degree classification judgment standard of the damaged building. In the evaluation of resonance by constant tremor, ground analysis by Shake is performed, seismic motion in the surface layer ground is created, transfer function and H/V spectrum are obtained, and resonance is evaluated from the ground predominant period and the natural period of the building. In addition, noise removal such as smoothing and low-pass filter is also processed on the server side for the measured value data. On the other hand, in PML, the damage rate of the building is calculated by obtaining the inter-story deformation angle from the bilinear model of the building consisting of the stochastic seismic velocity and the base shear coefficient of the building.

The operation is to receive the measurement result of constant tremor from the terminal, execute the analysis, and then save the data in SQL. In addition, considering the changes in the analysis method in the future, it is a prerequisite that the input data is appropriately stored in the server and that the data stored in SQL is specified or provided in a form that is easy to see for the majority. The hardware or software environment of the server is selected in all aspects such as capacity, speed, stability, etc., to meet these assumptions, but by consolidating analysis or data storage, etc., in one place, data falsification can be done. There is no room stable results can be provided, and there is the merit that the burden of analysis on the terminal can be removed. To execute these tasks, the server development environment utilizes **Table 2**.

Function, environment	Explanation
Windows server	Operation system
Visual studio	C++,C#, compiler
Google geocording	Latitude and longitude from the address
Firebase	Communication between server and terminal
PHP	Dialogue with the database
JSHIS	Internet site
MYSQL	database
SHAKE	Ground analysis

Table 2.
Server development environment.

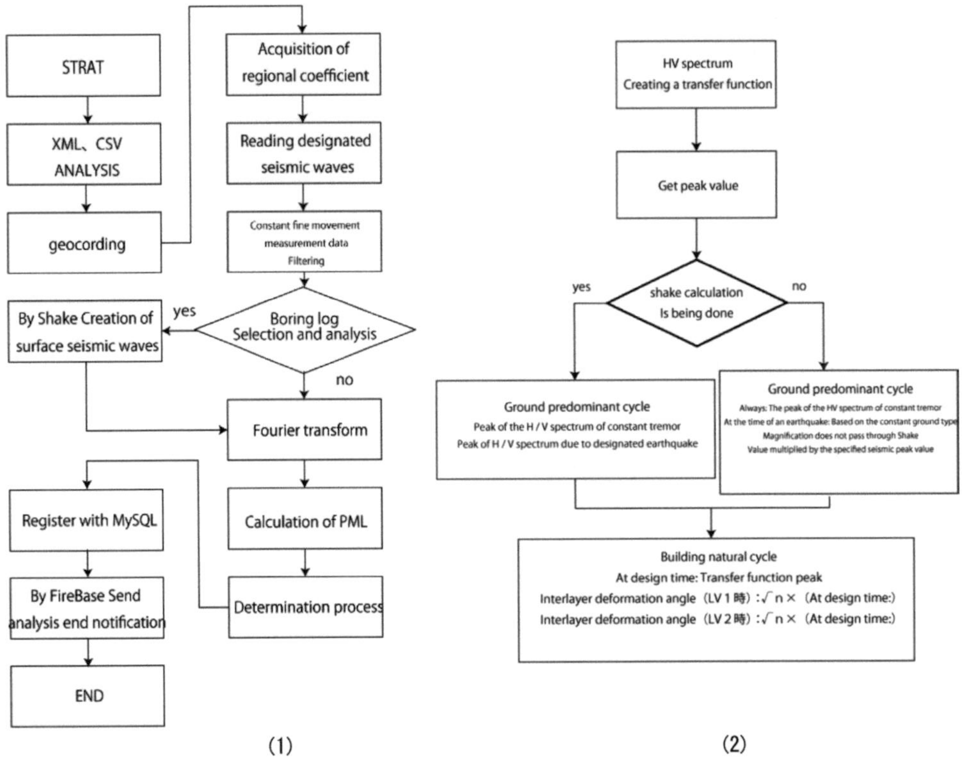

Figure 4.
Overall analysis flow (1) and judgment flow (2).

3.2 Selection of boring logs and analysis by Shake

In this research, XML data of boring log is obtained from the national land information retrieval site "Kunijiban" [2]. The coordinate values and data in the data are saved in SQL in advance, and the closest ground data is acquired from the data corresponding to the search near the area of 200 m, and the data is analyzed. The data of the boring log obtained by this is automatically generated according to the input specifications for Shake, and the surface seismic motion is generated. In addition, the shear wave velocity is input at the time of input, but in this study, the commonly used formula shown in the road bridge specification was used to estimate Vs (**Figure 4**).

3.3 Creation of H/V spectrum and transfer function

The H/V spectrum can be obtained by using the amplitude spectrum for each direction obtained by the Fourier transform in the following equation.

The period in which the H/V spectral ratio shows the maximum value is the natural period of the surface layer ground, and the maximum value is the estimated value of the amplification factor. In this study, the period of the maximum value of the H/V spectrum is treated as the constant ground predominant period of the surface layer. In addition, the transfer function is obtained by the following equation when the dynamic interaction between the ground and the structure is taken into consideration. The period of the maximum value of the transfer function is the building natural period:

$$S_i = \sqrt{\frac{X_i^2 + Y_i^2}{Z_i^2}} \tag{1}$$

Si : H/V spectral ratio at i Xi : Spectrum in the X direction at i
Yi : Y-direction spectrum at i Zi : Z-direction spectrum at i

$$Tf_i = Bi / Gi \tag{2}$$

Tfi : Transfer function at i
Bi : Fourier spectrum of the building at i
Gi : Fourier spectrum of the ground at i

3.4 Natural period of buildings and ground predominance period at the time of earthquake

The range in which the natural period of the building transitions can be calculated as the resonance range by specifying the interlayer deformation angle for the assumption of the interlayer deformation at all times and the assumption of the interlayer deformation angle at the time of an earthquake. **Table 3** shows this study.

It is the specification of the specified interlayer deformation angle. In addition, if the calculation of the ground predominant period Shake at the time of an earthquake is available, the predominant period is always obtained from the specified seismic wave by the same method, but if the acquisition of the boring columnar map fails, the ground is always obtained. The multiplication value α is defined from the value of the predominant cycle and is defined as the ground predominant cycle at the time of an earthquake (**Table 4**). Each cycle is considered in a band shape, and the future emergency risk judgment is predicted from the occupancy rate of each area (**Figure 5**):

$$T_B = \sqrt{\frac{\Theta_A}{\Theta_B}} \times T_A \tag{3}$$

Θ_A: Interlayer deformation angle A Θ_B: Interlayer deformation angle B
T_A: Period at inter-story deformation angle A T_B: Period at inter-story deformation angle B

	Steel	Reinforced concrete	Wood
At design time	1/300	1/400	1/200
During an earthquake (LV1)	1/200	1/250	1/120
During an earthquake (LV2)	1/100	1/125	1/60

Table 3.
Setting the interlayer deformation angle between design and earthquake.

Ground type	Natural period	α
First-class ground	~0.4	1.0
Second-class ground	0.4~0.6	1.5
Third-class ground	0.6~	2.0

Table 4.
Setting of ground type and multiplication value α.

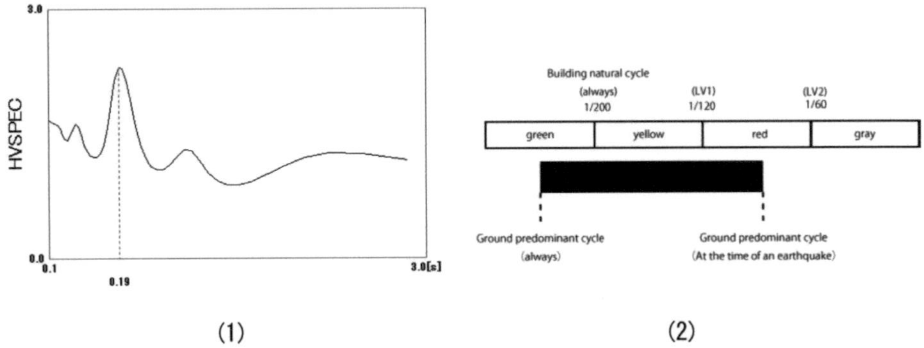

Figure 5.
H/V spectral ratio (1) and band determination (2).

3.5 Analysis by PML

In the analysis by PML, the probabilistic seismic velocity and s related to the ground [5]. The propagation speed, amplification factor, etc., are obtained from JSHIS, and the building is bilinear.

Apply to the model and calculate its damage rate. Since the PML value at the time of an earthquake obtained here is an index of a general building, considering the

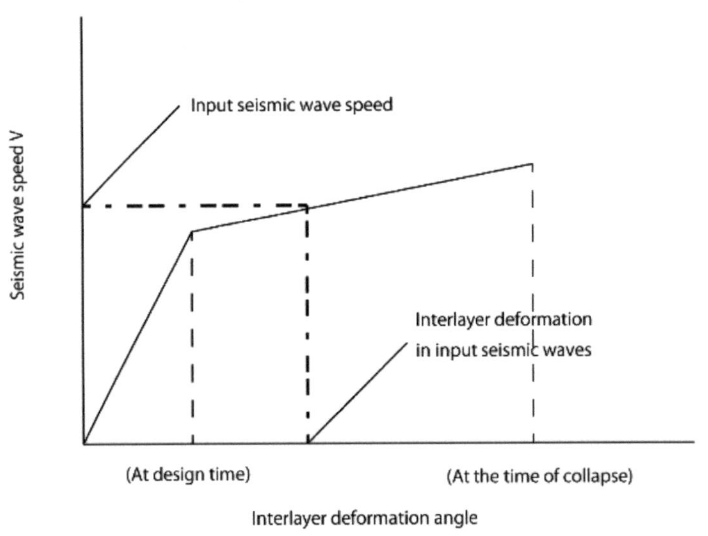

Figure 6.
Bilinear model at the time of PML calculation.

Judgment	PML value (%)
green	~15
yellow	15~40
red	40~70
gray	70~

Table 5.
Judgment as PML value.

validity of the measurement target, the abbreviation formula of the building natural period that is generally used and the building natural period obtained by constant tremor are used. By multiplying the ratio of, the PML value of the seismic motion considering the building to be measured was calculated (**Figure 6**). Additionally, the defined criteria based on the obtained PML values are presented in **Table 5**.

4. Discussion

4.1 Aggregate results

Distribute the application on Google Play. As a result, there are collect the results of 52 measurements. As a tendency of the measurement target. There were 26 cases of reinforced concrete, 18 cases of steel frame, and 7 cases of wooden construction. Most buildings are 1 to 15 stories high and are located in metropolitan areas.

4.2 Features by category

In terms of structural types, among the 18 cases of steel frame structures, 11% were classified as red, 50% as yellow, and 39% as green. Among the 26 cases focused on reinforced concrete structures, 4% were classified as red, 58% as yellow, and 38% as green. As for the wooden structures, out of the 7 cases, 29% were classified as red, 14% as yellow, 29% as green, and 29% as collapse.

Furthermore, focusing on the aggregation based on the ratio between the probabilistic seismic velocity from JSHIS and the S-wave velocity (Vs), when dividing it by the average, the results showed that for cases below the average, 68% were classified as green and 32% as yellow. On the other hand, for cases above the average, 70% were classified as yellow, 22% as red, and 8% as collapse, confirming the dependence on ground characteristics (**Table 6**).

Result	Location	V/Vs
Yellow	Kameari, Katsushika-ku, Tokyo	0.83
Yellow	Shinjuku, Katsushika-ku, Tokyo	1.00
Gray	Nishi-Nippori, Arakawa-ku, Tokyo	0.99
Yellow	Hibino, Mihama-ku, Chiba City, Chiba Prefecture	0.99
Red	Nogizaki, Moriya, Ibaraki	0.95
Red	Kagurazaka, Shinjuku-ku, Tokyo	0.67
Yellow	Oi, Kashiwa, Chiba	0.83

Result	Location	V/Vs
Yellow	Ueno, Taito-ku, Tokyo	1.01
Yellow	Ueno, Taito-ku, Tokyo	1.00
Red	Nishigotanda, Shinagawa-ku, Tokyo	1.00
Yellow	Kumotoji Tanida, Kuki City, Saitama	0.76
Gray	Saitama Prefecture Kita-Katsushika-gun Sugito Town Koya Tainan	0.82
Red	Kamitakano, Saitama	0.81
Red	Saitama Prefecture Minami Saitama District Miyashiro Town Gakuendai	0.81
Yellow	Saitama Prefecture Kita-Katsushika-gun Sugito Town Koya Taito	0.81
Yellow	Kameari, Katsushika-ku, Tokyo	0.83
Yellow	Tabata Shinmachi, Kita-ku, Tokyo	1.02
Yellow	Nishi-Nippori, Arakawa-ku, Tokyo	0.98
Yellow	Kanamachi, Katsushika-ku, Tokyo	1.02
Yellow	Kanamachi, Katsushika-ku, Tokyo	1.02
Yellow	Kanamachi, Katsushika-ku, Tokyo	0.73
Yellow	Toganecho, Katsushika-ku, Tokyo	1.01
Yellow	Senjujucho, Adachi-ku, Tokyo	1.00
Result	location	V/Vs
Yellow	Tomita, Oamishirasato, Chiba	0.43
Yellow	Kasugacho, Nerima-ku, Tokyo	0.39
Green	Aoba-ku, Sendai, Miyagi	0.12
Green	Hojo Higashi, Chuo-ku, Sapporo, Hokkaido	0.10
Green	Senju, Adachi-ku, Tokyo	0.33
Green	Yagiyamamotocho, Taihaku Ward, Sendai City, Miyagi Prefecture	0.09
Green	Kamihatagi, Hokota, Ibaraki	0.38
Yellow	Kashiwa, Kashiwa, Chiba	0.50
Green	Kashiwa, Kashiwa, Chiba	0.41
Green	Kashiwa, Kashiwa, Chiba	0.41
Green	Kashiwa, Kashiwa, Chiba	0.50
Green	Nishi-Nippori, Arakawa-ku, Tokyo	0.33
Green	Yaesu, Chuo-ku, Tokyo	0.34
Green	Japan Bridge, Chuo-ku, Tokyo	0.34
Green	Komagome, Toshima-ku, Tokyo	0.37
Green	Nishi-Ikebukuro, Toshima-ku, Tokyo	0.41
Green	Nishi-Ikebukuro, Toshima-ku, Tokyo	0.41
Green	Suehiro, Okegawa, Saitama	0.37
Green	Senju, Adachi-ku, Tokyo	0.33
Green	Senju, Adachi-ku, Tokyo	0.33
Yellow	Negishi, Taito-ku. Tokyo	0.34
Yellow	Ueno Sakuragi, Taito-ku, Tokyo	0.52

Result	Location	V/Vs
Yellow	Takada, Toshima-ku, Tokyo	0.58
Yellow	Takadanobaba, Shinjuku-ku, Tokyo	0.58
Yellow	Nakamachi, Kodaira City, Tokyo	0.34
Green	Matsudo, Matsudo, Chiba	0.34
Green	Kashiwa, Kashiwa, Chiba	0.41
Yellow	Kashiwa, Kashiwa, Chiba	0.50
Gray	Senju, Adachi-ku, Tokyo	0.33

Table 6.
Judgment value of aggregated data and classification by V/Vs.

5. Conclusion and recommendation

The summary of this study is shown below.

- Shows the specific procedure for configuring the risk assessment prediction system and operates it.
- The app was actually published to show the usefulness of the app.
- Land dependence was observed in comparison with the values obtained from the hazard MAP.
- It is difficult to build data covering the entire country, like the scattered data of bowling pin diagrams. However, in Japan, it is possible to develop such an application because seismic information covering the entire country is disclosed based on geological layers and past earthquakes, as seen on the JSHIS website.

Author details

Itsuro Yoshizawa* and Osamu Takahashi
Tokyo University of Science, Tokyo, Japan

*Address all correspondence to: 4122705@ed.tus.ac.jp

References

[1] Ministry of Land, Infrastructure, Transport and Tourism of Japan. Seismic Damage Classification Criteria for Disaster-affected Buildings and Restoration Technical Guidelines, Revised Edition. 2015

[2] Ministry of Land, kunijiban [Internet]. 2023 . Available from: https://www.kunijiban.pwri.go.jp/jp/

[3] Research Institute for Earth Science and Disaster Resilience. J-SHIS [Internet]. 2023. Available from: http://www.j-shis.bosai.go.jp/

[4] Development and Research of an Emergency Hazard Assessment System Utilizing Continuous Microtremor[thesis].Tokyo: Tokyo University of Science

[5] A Study on the Cost-effectiveness of Comprehensive Earthquake Risk Assessment Considering Structural Systems, Construction Sites, and Applications as Variables[thesis].Tokyo: Tokyo University of Science

Chapter 3

Perspective Chapter: Recent Advancements in the Management of Construction Risks

Naimah Muhammed-Yakubu

Abstract

The UK zero-harm agenda's catchphrase, "One Death is Too Many," demonstrates that no accident on construction sites is acceptable. Risks associated with construction can be reduced by combining cultural, social, and technological factors. While much work has been done from social and cultural perspectives, the technological aspect, particularly the use of digital technologies, has been minimal. To address this, it is necessary to investigate how social-technical systems interact and their effects on risks on construction sites. Although evidence indicates that effective digitization is required to achieve the zero-harm target and best practices on construction sites, current approaches focus on the socio-cultural aspect of risk management. In order to enforce risk management on construction sites, this paper addresses the digital visualisation of risk management by identifying risks and unsafe site practices. It suggests new approaches for utilising contemporary technologies to reduce risks on construction sites.

Keywords: UK zero-harm agenda, management of construction risks, social-technical systems, socio-cultural aspect of risk management, unsafe site practices, digital visualisation of risk management

1. Introduction

Every construction project, regardless of its size, complexity, nature, or location, has inherent risks that persist throughout its life cycle. All projects are susceptible to some form of risk, and failure to manage such risks effectively could be detrimental to the project's goals [1]. Risks are events that are uncertain and could potentially lead to negative consequences that may affect the project's success [2]. Although they can be minimised, managed, or eliminated, they cannot be disregarded/ignored. Risks are typically categorised as either generic or project based [3, 4], but their nature, form, or impact on a specific project may not be apparent until later during the project's execution. While a project progresses through its various stages, new risks may emerge, change, or even reduce existing ones. Regardless of the cause, it is important to take a proactive approach to mitigate these risks to prevent resource wastage, increased project costs, reduced productivity, and longer project completion times. The construction industry places a significant emphasis on identifying and managing risks throughout the construction process [5] because the process is susceptible to numerous risks due to various factors. These factors may include industry fragmentation, varying conditions of construction sites, intricate procedures, lengthy project timelines, and relationships among project participants [6, 7].

These factors are further compounded by the untimely identification of project risks which makes the sector keen on recent developments in risk management. Literature indicates that the management of risks in construction projects has not been properly handled, leading to suboptimal results [1]. Recent studies have however identified several trends and innovations in risk management, such as joint risk management [8], the use of Building Information Modelling (BIM) [9, 10], sustainable risks management [11], and cyber security policies [12]. The focus on assessing and managing risks has also led to the development of new techniques and protocols to prevent and mitigate risks on construction sites. Research has shown that effective risk management is crucial for project performance. As such, there is a growing awareness of the need for proper risk management procedures. To minimise errors, disruptions, and other undesirable events on-site, constant efforts are being made to improve risk management methods and procedures on construction sites [13]. Managing risks involves planning, monitoring, and controlling measures to prevent and mitigate potential harm to projects. The first and most critical step in risk management process is identifying potential risks. Without this identification, it is impossible to proceed with the subsequent stages of risk management [5].

Despite many efforts to identify risks and unsafe practices in the literature, current reviews on the topic are limited. Additionally, the existing analyses do not investigate unsafe site practices on construction sites, socio-technical interactions, and their effects on construction risks to address the digital visualisation of risk management. Only a few studies have reviewed construction risks from cultural and social perspectives, while technical perspectives, especially the use of contemporary technologies, have been minimal. Therefore, it is necessary to investigate how social-technical systems interact and how this affects risk on construction sites. The study aims to address the digital visualisation of risk management by identifying risks and unsafe site practices to enforce effective risk management on construction sites. The study adopts review of existing literature on the topic to identify the sources of risks to enable effective planning, monitoring, controlling, and management of projects. This aim would be achieved through the following objectives: identify risks and unsafe site practices on construction sites, investigate how socio-technical system interact and its effect on construction risks and address the digital visualisation of risks management on construction sites.

Section 2 of this study provides an in-depth review of literature on construction risk management, covering such topics as identifying potential risks and unsafe practices on construction sites. Section 3 outlines the methodology used in the study. Section 4 discusses socio-cultural and socio-technical considerations of risk management, including how they interact and their effect on construction risk. Furthermore, the section examines the use of digital technologies visualisation in risk management. The results and discussion are presented in Section 5, and the final section concludes the study.

2. Construction risks management: literature search

In the construction industry, risk management involves identifying and assessing unforeseen occurrences that may arise during project construction. The aim is to use tactics and procedures to minimise the identified risks to a manageable level to ensure project success. Risk management takes different forms depending on the situation, and it is a detailed and rational approach to defining, evaluating, and addressing risks in construction projects to achieve the project objectives [14]. Risk management is a methodical process that focuses on recognising, evaluating, and responding to the identified. The management of construction risks has traditionally centred on financial, project performance, productivity, and time aspects [15–18]. However, managing construction risks is a critical issue that requires attention as it relates to all events that can potentially impact the outcome and success of projects [19]. Thus, it should not be limited to financial or project performance, productivity, and time aspects. A well-planned project ensures an effective risk management [19]. This ensures that the project runs smoothly from the initial stages throughout its execution stage.

Furthermore, construction sites are prone to development and instability, resulting in hazardous environments and complex projects, which can lead to complications in projects operations. Therefore, it is necessary to identify all possible risks to minimise damage. While each project is unique, effective risk management during the conceptualization, execution, and evaluation phases is crucial. This highlights the importance of risk management throughout the entire project process. To ensure project success, measures must be taken to identify risks and implement effective risk management strategies [15]. Monitoring and adapting to changing circumstances is also crucial. Only then can strategies be developed to minimise risks and achieve project success.

The following sections discuss some of the techniques adopted in the reviewed literature to identify construction risks, including additional approaches identified from the results of the review, and then discussions on unsafe site practices on construction sites.

2.1 Identification of construction risks

Identifying the sources of risk is a crucial step and is considered the first and most important in risk management process [19], but this process can only be performed on identified [5]. This allows all project parties to pinpoint specific instances of risks, assess their potential impact, and analyse them to develop appropriate measures for mitigating their effects. An unidentified risk cannot be managed. A comprehensive risk identification process ensures an effective risk management process [20]. Equally, the identification of risks can reveal possible opportunities from the identified risks. However, project executors often concentrate more on the identification and management of the negative risks and overlook the positive aspect/opportunities that could be derived from the identified risks [21]. Identification of risks is a continuous process, that often require rigorous tasks and constant monitoring throughout the project life cycle because existing risks may cease to exist, and new risks emerged [2].

Many sources suggest that trying to identify every possible risk in a construction project is a difficult task and can have negative effects [22, 23]. Instead, it is recommended to focus on the most common and impactful risks. Identifying risks is an ongoing process throughout the project's lifespan, and it requires input from all stakeholders, including external risk management experts [2]. Researchers believe that involving stakeholders can instil a sense of responsibility for the risks identified in the project, encouraging them to take ownership and develop effective strategies for eliminating or mitigating the identified risks. Furthermore, external stakeholders and experts can offer valuable information, such as standard risk checklists, risks identified from previous project, risk structures and registers. This information can aid in identifying risks and contribute to the success of the project.

The literature highlights different methods for identifying risks, such as the SWOT analysis approach (strengths, opportunity, weaknesses, opportunities, and threats) [24–26], information-gathering [20, 27, 28], document reviews [20, 27, 29, 30], risk workshops (Top-Down method) [31, 32], interviewing (Top-Down method) [20, 33, 34], identification based on possible risks scenarios (Top-Down or Bottom-Up) [35, 36], identification based on the root cause of the risk [37, 38], and surveys or questionnaire (Bottom-Up) [28, 39]. **Table 1** below provides additional information about the different approaches for identifying risks and **Figure 1** depicts the visual representations of the procedures for identifying construction risks.

- Checklists for assessing risks
- Portfolio of previous projects
- Deficiency in construction project performance
- Identifying risks with PESTLE analysis

2.1.1 Checklists for assessing risks

It is important to have a risk assessment checklist to create an effective plan for identifying, estimating, assessing, prioritising, and managing risks. This checklist ensures that all aspects of the project are evaluated thoroughly, and all relevant information is incorporated into the risk management plan. This method is a critical step in identifying, managing, and accessing risks. Although the implementation may vary depending on the scope and type of project. However, a general checklist for assessing risks that can be applied to all projects is provided below.

a. **Funding:** When considering a project's scope, it is important to analyse the different types of risks that could impact funding and consider the areas where funding risks could arise. Without sufficient cash flow or funding, progress in the project cannot be made [40]. Therefore, it is necessary to identify any factors that might affect cash flow. By identifying and mitigating potential funding risks, the project's expenses can be better managed and kept within budget.

b. **Project timeline:** To determine if the estimated time for a project is correct, the schedule of work for the entire project should be carefully analysed. This includes a critical examination of daily operations and operational timelines. Additionally, external factors like weather, human error, incoming revenue, and involvement of external parties should be considered and included in the evaluation process.

c. **Project resources:** The project's resources must include all necessary items and materials, including both internal and external resources like employees and suppliers [1]. All project employees should be trained to handle their tasks and responsibilities, and new responsibilities should also come with proper training. It is important to consider what materials or assets will be needed for each

Methods	Benefits	Challenges
SWOT (strengths, opportunity, weaknesses, opportunities, and threats) analysis approach [24–26]	This method facilitates convenient access to information for planning, decision-making, and risk management by continuously updating the SWOT analysis grid template to assist the project team in identifying project risks. The tool aids in determining which risks should be given priority to achieve the expected project objectives. Additionally, it provides a backup plan to ensure that the project stays on track within the minimum cost limitations	Constant review, analysis, and updating of information are a challenging procedure in this method
Information-gathering [20, 27, 28]	Collecting data is a vital step in identifying potential risks and implementing effective administration strategies while ensuring the protection of valuable assets. It is crucial in revealing other potential risks that may exist	This method often fails to give a holistic overview of risk in the identification process due to different project structures
Document reviews [20, 27, 29, 30]	This technique is the most cost and time-effective approach due to its independent nature and simple process. The review method helps to identify the construction company's strengths and weaknesses while also providing insight into the project's history and philosophy	When using this method to identify risks, it's important to assess its quality alongside the results of other methods like interviews and questionnaires
Risk workshops (Top-Down method) [31, 32]	Risks workshops can assist in identifying various types of risks such as new, evolving, or emerging risks. These workshops can also help in discovering connections between risks, identifying the appropriate project teams to take ownership of these risks, and establishing a culture of risk management within the industry. Typically, workshops focus on identifying top risks, but are suitable for small to medium-sized projects	This may not be appropriate for large-scale projects. It can be challenging to identify the participants for the workshop, as many teams tend to work independently and may not have a comprehensive understanding of the overall risks involved in the project. While they may be experts in their respective areas, their contributions may not extend to the entire project. Additionally, it can be difficult to establish clear goals for the workshop, and synthesising the comments and defining the risks in a common language can be a daunting task. Compiling a comprehensive list of high-level risks can also pose a challenge
Interviewing (Top-Down method) [20, 34]	The process involves individual meetings between project team leaders to share their opinions on major project risks. While additional individuals may be interviewed later, this method has proven to be effective in uncovering sensitive risk issues. Each participant can freely express their thoughts in a private setting, leading to more open and honest discussions. This one-on-one approach also helps gain crucial support for the risk management process from top-level managers, as it allows them to ask questions privately	The biggest challenges in using the interview method is interpreting the results, which involves concerns and classification based on the organization or project. However, this challenge can be solved through pre-planning and scheduling to ensure that both the interviewer and interviewee have a clear understanding, ambiguity is corrected, and any miscommunication is resolved

Methods	Benefits	Challenges
Identification based on possible risks scenarios (Top-Down or Bottom-Up) approach [35, 36]	This approach is most effective for identifying risks that have a low probability of occurring but could have a significant impact on the project if they do. It can be especially helpful for leaders who are looking to make strategic decisions and want a better understanding of potential risks. Scenario risk identification can help uncover new risks and improve existing mitigation measures. It can also help identify blind spots, biases, and interdependencies. This approach is particularly useful when there is a major change underway that could have a significant impact on the project	Identifying risks by this method can be challenging due to scheduling constraints and participants may take issue with the "what if" nature of the questions, as they are based on their responses. Remaining calm, flexible, and focusing on high-impact risks can help rectify the situation
Identification based on the root cause of the risks (Bottom-Up) [37, 38]	This method involves identifying the root cause of issues to discover project risks. It is distinguished from other methods by the fact that it begins with a known problem and then examines it closely to gain understanding, rather than simply asking about potential risks. Additionally, it can uncover opportunities to create additional value and identify new risks. Both employees and operational level managers are involved in this process, which can reveal emerging risks that may not be apparent to others involved in the project	This method delves into the processes in detail to ensure that the results of root cause analyses are thoroughly understood by various departments. This process takes longer than other methods as it is crucial for accurate identification
Surveys or questionnaire (Bottom-Up) [28, 39]	This "Bottom-Up" method encourages lower-level employees to speak up and share their opinions and concerns regarding potential risks without fear. This approach involves distributing surveys to multiple individuals working on the project to identify various risks throughout the project. It is particularly useful when traditional methods, such as interviews and workshops, are not feasible due to the geographical location of stakeholders. Typically, this method focuses on middle managers, teams, and other support staff involved in the project. The survey questions are open-ended, allowing employees to express their thoughts in their own words about the most significant risks they perceive. Moreover, participants sometimes rate how well management is addressing those risks. This approach provides a unique perspective on project risks from the employees' viewpoint	This method requires handling a large volume of information which can be challenging, and without the proper procedures and tools to capture, analyse, and report responses, desired results may not be achieved

Table 1.
Risks identification methods.

task in the project, and how they will be obtained and maintained sustainably. However, there may be instances where resources are unavailable or do not meet expectations. To prepare for such scenarios, a plan to mitigate the risk of resource shortage or delay should be included in the project checklist.

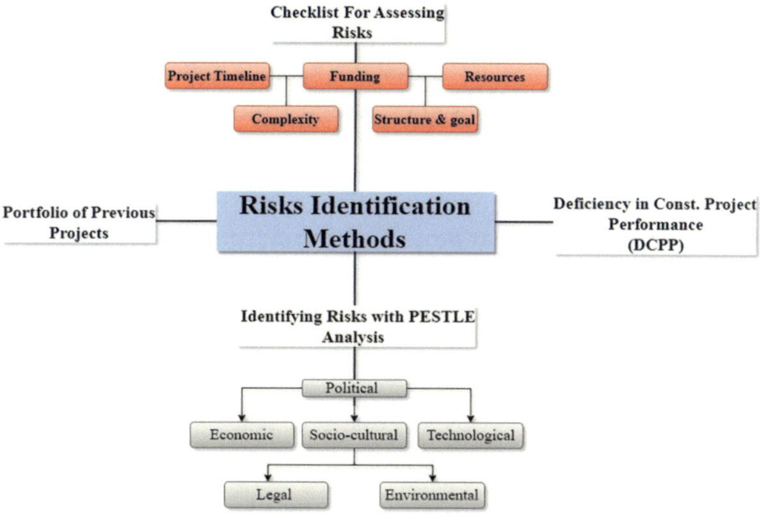

Figure 1.
Approaches for identifying risks on construction sites.

d. **Project complexity**: When evaluating the complexity of a project, it is important to consider the various stakeholders involved. More complex projects tend to involve a greater number of stakeholders, which means more consultation and management is required. To ensure the project runs smoothly, regardless of size/complexity, it is crucial to understand the full responsibilities of each stakeholder and how their involvement could potentially impact the project [41]. This includes identifying any potential risks and developing strategies to manage them effectively.

e. **Project structure and goal**: Finally, it is important to identify and address the potential risks of the project, as well as the objectives of the project which may include those related to the task, operational area, and overall goals [42]. By doing so, any potential obstacles that could impact the success of the project can be identified and evaluated before they become a problem. To accomplish this, it is essential to ensure that all team members are aligned with the project goals and that those goals are both realistic and meaningful.

2.1.2 Portfolio of previous projects

To effectively manage risks, it is crucial to first identify and discover them [43]. One way to accomplish this is by considering risks from previous projects. The impact of risks that have been identified but not addressed in previous projects normally have a lasting impact on project managers and influence their decisions in future projects [43]. However, this information may not be readily available to project managers, and their lack of knowledge may contribute to the issue of unmanaged risks. That is why having a portfolio of previous projects can be helpful. This is essentially a collection of past projects/programs designed to manage important information across project implementations, providing guidance and oversight for future projects [44]. Project portfolios are essential in achieving goals and implementing strategies. They provide guidance in making informed decisions regarding resource allocation, forecasting project performance, and identifying risks, ultimately leading to project alignment, progress, and increased productivity. Risk management greatly benefits from project portfolios as they ensure strategic alignment, reduce inefficiency, evaluate projects, and promote diversification.

When analysing data from past projects, it can be valuable to utilise corporate knowledge. However, this technique may not be feasible if the organisation has not previously undertaken a similar project or if data from a past project were not recorded. In these instances, database systems that track and report project progress can be helpful in identifying potential risks. Other useful tools include record of lessons learned from past projects and evaluations of previous projects. However, it is important to note that these method/systems may be limited by the quality and relevance of the available data.

2.1.3 Deficiency in construction project performance (DCPP)

Deficiency in construction project performance (DCPP) can either be minor or major and may cause damage to property or people. Regardless of the severity, it is often not discovered until after the project is completed, making it an expensive risk and hard to fix. Poor design, materials, workmanship, structural failure, and financial risks all contribute to DCPP. This can lead to serious structural defects that fall short of the owner's expectations, but the severity of these defects can vary greatly. According to [45] these defects can be either Patent, meaning they are readily known, or Latent, meaning they are not easily observable. Latent risks are more problematic as they are not easily detected even with constant supervision. Patents, on the other hand, are visible and can be easily fixed. They are referred to as surface-level risks, while latent risks are known as below-surface risks due to their less visible/less obvious nature.

When a design professional fails to produce accurate and well-organised construction documents or creates out-of-scope designs, *poor design* may result. This can also happen due to errors or omissions. When errors occur, the professional must redesign and replace the erroneous design. Omissions, on the other hand, can be remedied by adding a "*change order*" to the contractor's scope of work. A change of order is a document used to modify an already completed construction project [46]. It includes details about the changes in scope of work, cost, and schedule that were not initially included or corrected in the initial project. Completing a change of order correctly reduces risk, increases the chances of approval, and helps contractors fulfil their financial obligations faster [47]. However, if a change of order is not completed properly, it can affect the contractor's liability and put their financial obligation at risk.

When construction *materials* are inadequate or damaged, it can lead to deficiencies in the final product. Often, the issue is not discovered until the material has already been incorporated into the project. This can result in costly risks, including damage to the structure and additional expenses for labour and materials. *Workmanship*, which refers to the quality of the human labour involved in the construction process, is another important factor. Workmanship issues can range from minor surface flaws to major structural problems, and it can be difficult to determine who/what failed to meet the property development standards [48].

If the project participants fail to meet their contractual obligations according to the required standards for developing a structure, it may lead to a *failure in the structure's overall structure (structural failure)*. This happens when the approved design and contract documents are not followed. For instance, not familiarising oneself with the site's local conditions as per the contract or regular visitation, not coordinating onsite activities by reviewing the contract documents constantly, and not ensuring that the work performed follows the acceptable standards of workmanship [49]. These are considered the general conditions that every contractor should adhere to while developing any structure. They form the backbone of the entire contract and govern the project's rights, responsibilities, duties, privileges, and rules.

One unique aspect of construction projects is that *financial risks* are the first contributor to the delays, changes, and quality problems. With so many businesses involved and a lack of visibility into who is responsible for each task, financial issues can become more complicated. Construction projects carry significant financial risk, including budget and cash flow concerns, which can contribute to DCPP. Additionally, the construction industry is credit-heavy, with all materials and labour furnished before payment is received. This means that participating teams often face challenges in getting paid, waiting longer than in other industries to complete time- and information-intensive payment applications. While waiting for payment, they must find ways to keep cash flowing and cover upfront expenses. Sometimes, even when payment is received, it may not cover all expenses, leading to inefficiencies in project performance, budget, cost overrun, and other associated financial risks. Contributing factors to these payment challenges include complicated payment structures, high failure rates, hidden parties, confusion or complex contractual issues, and other factors that can result in financial damages/risks [50].

2.1.4 Identifying risks with PESTLE analysis

PESTLE analysis is a useful tool for examining and identifying risks in projects. By looking at risks from six different perspectives—Political, Economic, Sociocultural, Technological, Legal, and Environmental—it provides a comprehensive understanding of potential threats and opportunities. This analytical tool is commonly used in Strategic and Operations Management (SOM) to aid in the process of risk management in businesses. When used in conjunction with SWOT analysis, PESTLE helps to identify both internal and external risks [51]. Internal risks are easier to identify as past data from similar projects or portfolios of previous projects are available, while external risks are not easily known and are beyond the control of the company. Thus, employee lacking and leaving the construction projects vulnerable to major failure [51].

2.1.4.1 Political risks

The construction industry, like any other organisation, operates within a political environment that presents certain risks. Political risks refer to situations that are dependent on political and regulatory factors, as well as the overall stability of the country where construction is taking place [52]. It is important not to ignore political risks in construction because they can lead to uncertainty in the political landscape of construction projects [53]. These risks can include breaches of contracts, terrorist attacks, and wars that may impact projects [54]. Political risks can manifest in a variety of ways, such as changes in government laws, regulations, and policies that affect the project, political instability within the government, delays or refusals of project approval and permits by government departments, and the outbreak of hostilities such as wars, revolutions, riots, and terrorism [53]. Unlike other types of risks, political risks are more complex, unpredictable, and devastating because they fall outside the scope of normal project activities [55].

Political risks are a crucial part of risk management in construction projects. However, it is essential to recognise that project-level political risks can also impact overall business objectives like development and strategic decision-making [56]. While managing political risks, it is important to avoid overemphasising short-term project goals and neglecting corporate strategic objectives. Inappropriate resource allocation among projects, constraints of project resources, and a lack of risk management experience are some of the drawbacks that need to be addressed [53]. Therefore, it is crucial to consider political risks and link risk management strategies to the project's objectives based on sufficient resources and information. This is an important component of the decision-making process for the continuous improvement of the project.

2.1.4.2 Economic risks

Economic risks related to projects were discussed by various researchers including [57, 58]. These risks include inflation, fluctuating exchange rates, sudden changes in prices, tax rates and economic policies, as well as difficulties in financing the project. Liu et al. [59] also identified other economic risks such as unclear tax payment responsibilities, lack of partial payment provisions, and improper withholding of guarantees on advance payment. According to [51], inflation and sudden price changes are the most common economic risks in construction projects. El-Sayegh [60] emphasised the importance of accurately predicting inflation to determine future project costs. However, it is difficult to accurately forecast interest rates since they depend on the global economic climate, which may not always align with assumptions [61]. Therefore, it is recommended to consider adjusting project plans based on variable interest rates, even though they may not be accurate [60].

2.1.4.3 Socio-cultural risks

There are various social-cultural risks associated with construction projects, such as cultural and religious differences, lack of security on project sites, and social and cultural impacts on the community [51, 60]. Public objections to projects are also a concern. These risks are related to different aspects of people's lifestyles, demographics, educational levels, values, and ways of thinking [62]. They can affect any industry, but they play a crucial role in construction projects due to the diverse nationalities involved. For instance, Europeans tend to be confrontational and less respectful of hierarchy, while Asians tend to avoid conflict and prioritise seniority and hierarchy [63].

The way a project team cooperates, collaborates, and coordinates can be affected by socio-cultural factors. This, in turn, can impact the management style, speed, and processes of the project itself. This can help keep the project on budget and schedule, as noted by [64, 65]. Through these qualities' interaction with the project team's cooperation, collaboration and coordination, the socio-culture affects the management style, speed, and processes of the undertaken projects [65]. These qualities and differences thus need to be managed effectively especially where the project leaders/managers are foreigners to finish the projects within budget and time [64].

2.1.4.4 Technological risks

Project technical risks are associated with the technical aspects of a project, including design errors or changes, poor engineering, insufficient details in design specifications, and technology issues [51]. Additionally, research and development approaches, the impact of internet usage on project work, untested engineering techniques, delays due to lengthy design processes, complex design, poor buildability and constructability, and technology changes were identified by [1, 50] as forms of technological risks encountered in construction projects. Due to the dynamic nature of construction sites, technological risks can change frequently, leading to inadequate identification of risks, application, and mitigation measures on site.

2.1.4.5 Legal risks

Legal risks can arise due to various factors such as delays in resolving contractual disputes, poorly developed or unclear contract documents, changes in codes and regulations, conflicts in contract documents, and inappropriate distribution of responsibilities [52, 57]. Inadequate claim administration, rigorous competition at the tender stage, excessive contract variation, third-party liabilities, immature laws, and complexity in the legal environment can also contribute to legal risks. It is important to note that legal risks may vary depending on the project and country. For instance, negotiations and strategic decisions may be made to address risky site conditions in some countries, where risk allocation is done through contract clauses [66, 67].

2.1.4.6 Environmental risks

The construction industry faces a range of risks due to its dynamic and complex nature [68]. These risks include inadequate information about the site, inappropriate construction scheduling, client demands for changes, excessive noise pollution, and a lack of skilled staff to handle unexpected environmental conditions [69]. El-Sayegh and Mansour [52] identified additional environmental risks, such as uncertainty about underground conditions, the presence of archaeological artefacts, site accessibility, the unavailability of necessary infrastructure, security, and traffic conditions. Environmental issues such as failure to comply with environmental regulations and impact assessments can also lead to project suspensions and disruption of project goals [70]. Managing environmental risks is crucial for maintaining productivity and achieving project objectives in construction projects.

2.2 Unsafe site practices on construction sites

Construction projects are exposed to various risks that can both be anticipated and unexpected. These risks arise may be due to unsafe practices on construction sites. Such practices include unfavourable social and environmental conditions, physical hazards, congested workspaces, improper workstation layout, and movement of heavy equipment. Workers may also fail to take necessary precautions, such as wearing safety helmets and personal protective equipment (PPE) or maintaining proper working positions and conditions. Othman et al. [71] found that accidents on construction sites are often caused by a combination of the above issues. Meanwhile, Hon et al. [72] argue that accidents are not always due to construction operations failure, but to human error and unsafe site practices. Unsafe site practices refer to actions, intentional or unintentional, that may lead to accidents, injuries, or even fatalities among workers or materials [73]. Despite the existence of laws and regulations to ensure site safety, employers and employees do not always comply with them rigorously [43]. The consequences of these behaviours can be dire, and the lack of learning opportunities means that these practices tend to be repeated, unchecked.

Unsafe practices is consider a physical conditions or circumstances that could potentially cause accidents [73]. These practices are prevalent in construction sites and are responsible for many accidents. Researchers in various countries have studied the factors and causes of unsafe site practices. In Kuwait, inadequate safety procedures, improper materials, low maintenance, supervisory faults, and misplaced equipment were identified as key factors [74]. In the USA, Abdelhamid [75] classified unsafe site practices into human and physical factors. Human factors include incorrect working posture, failure to wear protective equipment, unauthorised use of equipment, unsafe operating speeds, poor mental health, removal of safety devices, and unsafe positions when operating equipment. Physical factors include the unsafe acts of others, disregard for prescribed procedures, accidents due to defects, unsuitable attire, unsafe environmental conditions, fire hazards, poor housekeeping, inadequate personnel assignment, and poorly guarded machinery. Toole [76] found that lack of training, inadequate enforcement of safety regulations, lack of safety equipment, unsafe construction methods, unsafe site conditions, failure to use provided safety equipment, poor safety attitudes, forgetfulness, and deviation from prescribed behaviours were responsible for unsafe site practices in the USA.

Additionally, construction site accidents have been attributed to unsafe practices in various studies conducted in China, Hong Kong, Australia, Turkey, Korea, and Greece [77–82]. These practices include inadequate training, inappropriate worker behaviour, poor safety awareness, reluctance to invest in safety measures, loss of control of tools or equipment, physical strain, reckless machine operation, lack of safety regulation enforcement, insufficient personal protective equipment, lack of innovative technology, collisions with objects, and poor information flow. Unsafe work processes, environmental conditions, and machinery/crane positioning and movement have also been identified as causes of accidents by some researchers [73, 80]. Thus, accidents on construction sites are largely caused by unsafe site practices. The construction project is a complicated and time-consuming process that involves many different specialised tasks typically carried out by humans. Studies have shown that construction risks and unsafe practices are caused by various factors, with human behaviours being the most significant contributing factor. However, four categories of unsafe practices are identified in this study: human factors, non-human factors, unsafe worker acts and actions, and unsafe environmental and site conditions.

3. Methodological approach to the study

Even though many studies have been conducted on risk identification and management on construction sites, none have examined how socio-technical systems of risk management affect construction projects. The study identifies risks and unsafe site practices through literature review, so that effective risk management can be enforced on construction sites through digital visualisation. Through the review, risk identification methods and unsafe site practices were examined qualitatively, and the results analysed. The identification techniques were discussed based on the views of various authors in the reviewed literature.

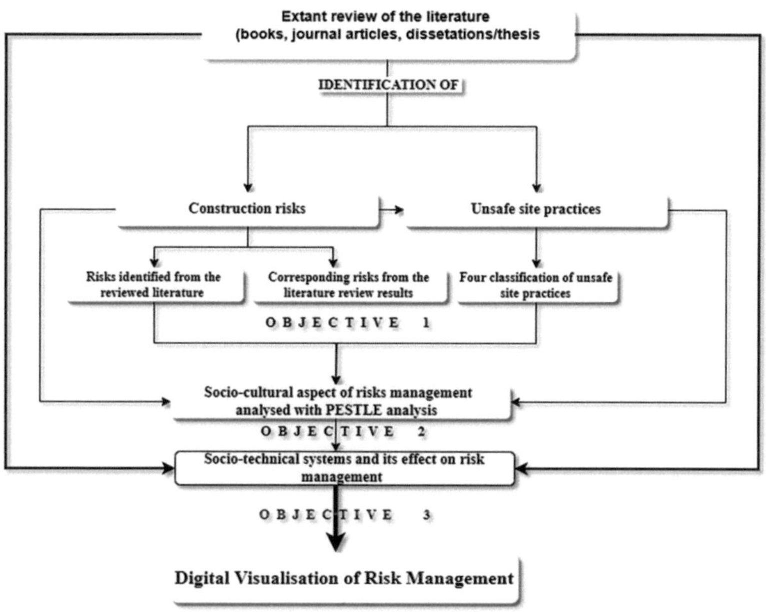

Figure 2.
The study's methodological approach.

The process of reviewing started with gathering relevant information from books, articles, and websites about risks and unsafe practices on construction sites. Through this search, the first objective was achieved, and the results of the literature review helped identify other methods addressing risks and unsafe site practices. From the identified methods, the fourth technique, PESTLE analysis, identified risks based on political, economic, socio-cultural, technological, legal, and environmental factors. The PESTLE analysis provided a powerful insight into risks from various angles and how some systems interact with risk management. Thus, achieving the second objective. The study also discussed the socio-cultural aspect of risk management based on PESTLE analysis classification including their interaction with socio-technical systems. This helps to achieve the third objective of addressing digital visualisation for risk management from socio-technical perspectives. **Figure 2** depicts the study's methodology approach.

4. Socio-cultural considerations of risk management

Evidence from studies indicate that socio-cultural considerations of construction risk management exist in different literature, as noted above in Section 2.1.4. PESTLE analysis revealed that socio-cultural factors related to risk management could affect construction projects in the society/organisation where they are carried out [83, 84]. From the review conducted, some studies identified socio-cultural related factors to risk management as depicted in **Table 2**.

Problems with land acquisition and compensation: According to [123], rapid urbanisation has created issues with land acquisition. He stated that the eight categories of risks associated with urbanisation are: infrastructure risk, population-structure risk, public-health risk, risk of conflicts of interest, risk associated with energy resources,

Socio-cultural considerations of construction risk management	Evidence from the literature
Problems with land acquisition and compensation	[85–89]
Public resistance to the project (Public concerns and social complaints)	[1, 90–92]
Diversities in social, cultural, and religious backgrounds	[93–95]
Crime and civic unrest	[96–101]
Conflicts at work and protests	[1, 62, 102–106]
Unfavourable public relationships	[107–110]
Degraded social environmental conditions	[111–114]
Social instability and/or societal conflict	[115–118]
Ineffective governance structure	[1, 119–122]
Disruptions to structure	[1, 123–126]

Table 2.
Socio-cultural considerations of construction risks management.

risk associated with environmental pollution, risk associated with conflicting conceptions of value, and risk associated with the differentiation of rural and urban areas along with the covert threat of social conflicts. In project development, land acquisition and compensation procedures are extremely challenging, and it has been attributed to rapid urbanisation in China by various researchers. These researchers noted that 65% of public disturbances in rural areas in China occurred because of land acquisition and compensation conflicts between 2004 and 2009 [123]. As a result, the current system for evaluating the risk of land acquisition is implemented by the Department of Land Management. This way, clarifying land risk evaluation targets and orientations improves the land acquisition risks assessment system and optimises land evaluation by making the assessment process more transparent and reasonable [123].

Public resistance to the project (public concerns and social complaints): [124] identified five factors responsible for Korean megaproject delays, time overruns, and cost overruns. These factors include poor project management and monitoring systems, conflicts between organisations, and strong public resistance. In addition to the causes common to Korean mega projects, they can also occur in other types of construction projects, but they tend to produce poorer results in mega projects compared to smaller ones [124]. Therefore, project size and complexity have an impact on public resistance.

Diversities in social, cultural, and religious backgrounds: Diversity comprises behaviours that consider our shared similarities and distinctions. As a result, every employee at a company has a distinct background that sets them apart from one another by virtue of their cultural, social, and beliefs. Diversity entails understanding of the people which involves human characteristics and knowledge; acknowledging each individual ethics, and enshrining relationships across all boundaries for people to work in tandem to eliminate all kinds of prejudice and bias [125]. Academically, diversity incorporates multiple aspects of humans such as their heritage, socioeconomic standing, ethnicity, speech, complexion, mindset, morals, convictions, and belief [126]. In addition, diversity is related to minorities comprising female gender, the disabled, and the elderly. Unmanaged diversity, however, may be a barrier to attaining organisational and project goals as it could decrease productivity and provide competitive advantages [127]. Thus, to effectively manage employees, project tasks, risks, and assets to increase productivity, diversity must be managed and valued.

Crime and civic unrest: In times of civil unrest, construction sites are susceptible to loss. A construction site can be a target for disruption for stored materials or partially constructed structures. This unrest increases the likelihood of property damage, crime, theft, fraud, vandalism, dishonest behaviour, and related incidents. These times can result in severe losses. For instance, police, firefighters, or emergency response may take longer to respond in theft, vandalism, and fire cases. Among the ways to handle this is to create a suitable plan through risk analysis and identification. Additionally, to ensure people's safety on and off the construction sites, identify known and anticipated risks, and develop safety protocols. This procedure identifies the parties responsible for implementing control measures to ensure the plan's effectiveness and the safety and well-being of those involved.

Conflicts and protests at work: Conflicts are inevitable in any construction setting. Conflict issues caused delays, interruptions, suspensions, or abandonments of projects [128]. In other words, identifying the conflicts helps to manage or eliminate the risks resulting from conflict and protests at construction work. Knowing the conflicting factors in advance makes it easier to deal with them.

Unfavourable public relations: An organisation's public relations is an analysis of trends, a prediction of their consequences, and a plan for implementing an action program that will benefit both the organisation and the public. When the intended goal of the planned program of actions is not accomplished, and the organisations and the public interest are not served, poor and unfavourable public relations result. Failure to define a good communication approach and stakeholders' information needs is a major cause of unfavourable public relations. Therefore, early identification of risks associated with lack of communication within project parties is necessary.

Degraded social environmental conditions: These include issues like land clearance, people's relocation, and health concerns from waste releases from construction projects. In view of this, it is standard practice to carry out an environmental impact assessment (EIA) in the project decision phase to consider social environmental issues for the end-users and the local community [129]. EIA evaluate the environmental effects of a planned operation [130]. It considers things like alternatives for the project and mitigation strategies if the construction were to proceed. An Environmental Impact Statement (EIS) or Environmental Statement (ES) summarises the findings of the activity. A successful EIA comprises individual and public assessments, and considers environmental factors like population, landscape, heritage, air, climate, soil, water, floral and other living creatures [131].

Social instability and/or societal conflict: Population growth and modern civilization result in social instability/societal conflicts [132]. It typically resulted from conflicts of interest between different project participants or stakeholders and is linked to societal disputes, violent opposition, and mass occurrences of public unrest [133]. Social instability cannot be completely eradicated, and poor management of it might jeopardise social harmony. However, it can be controlled to reduce the impact before it becomes a full-fledged unmanageable risk [134].

Ineffective governance structure: Governance structure is one of the main reasons for the challenges facing construction projects [135]. The governance structure technically enables independent individuals with different long-term interests and strategic objectives to cooperate towards achieving a common goal [136]. When this is unsuccessful, it develops into a problem and poses threats to the projects. Governance structure serves as a key instrument for managing risk in projects and sheds light on the interactions between various construction project participants [137].

Structural disruptions: Structural disruptions are one of the risks associated with normal construction activities. They can be caused by terrorism, political instability, economic disruptions, intentional agent actions, and human-centred problems like strikes [138]. Some instances of structural disruptions include the September 11, 2001, terrorist attack on the World Trade Centre and the August 14, 2003, blackout in the Northeastern United States. [138] identified four methods of managing disruptions to structure. They include defining the nature of the underlying hazard giving rise to the risk, quantifying the risk through an established risk assessment process (by defining the pathways through which such risks might occurred), effective risk management which requires that the approach utilised fit the features and needs of the decision environment, and appropriate management rules and actions to be integrated with ongoing risk assessment and project partner coordination.

4.1 Socio-technical considerations of risks management

The common practice of managing risks on construction sites suggests that the workers and construction personnel are not well-informed of potential risks, unsafe site practices and other safety-related issues. This limits the ability to actively make informed decisions about construction projects and managing the risks therein. Also, because information related to risk monitoring and control, such as risk identification, risk assessment, and other information on risks, is not publicly available to the concerned industry, various challenge occurs in the risk management process. In fact, many construction firms do not share risks information on project with their employees. This practice restricts important opportunities for learning about risks in the construction sector. Risks relating to employee and organisation are crucial in implementing risks management. The socio-technical aspect of risks management is required to address the dimensions of human, organisation, and technology and to take their interaction with one another into account [139]. Given this a priority can improve easy identification of potential risks and unsafe site practices on construction sites. Thus, ensuring that risks-related information is effectively and accurately delivered, understood, and well distributed among project participating teams.

Traditionally, the science of information delivery is traced back to humans which forms a significant part of the concept of the socio-technical aspect of construction risk management. Projects using digital technologies must be designed and optimised in a way that considers organisational structures, technology and humans to achieve risk-free outcomes [139] because a task is always implemented by the interaction of a human, technology, and organisation [140]. Additionally, [141] agrees to the understanding of a socio-technical perspective which involve human (collaboration, qualifications, work assignment, and structures), organisational (knowledge, organisation & processes, culture, and ethics), and technological innovations (IT systems, automation, data science and management). Hence, implementing socio-technical interactions demonstrates a degree of technology, of changes in organisational structure, and of changes in employee independence.

4.2 Digital visualisation of risks management on construction sites

The idea of digitalisation makes it possible to incorporate digital technologies into risk management. It provides a new level of organisation and management for the entire risk management process. Hirsch-Kreinsen et al. [142] defined digitalisation as the incorporation of digital technologies, the use of applications built upon these technologies, and their networking systems into the process of socio-economic change. This definition demonstrates that digitalisation encompasses significant economic, social, and technical changes. The term Volatility, Uncertainty, Complexity, and Ambiguity (VUCA), which describes the concept of digitalisation and societal change explains the dynamic nature and distinctive complexity of the construction industry/businesses. Consequently, the sector must become more responsive and adaptive to meet challenges that are increasingly difficult to anticipate in a rapidly evolving world of technology [139]. These changes are significant enough to justify construction companies developing visualised technological strategies to manage risks. For example, with computer vision, Internet of Things (IoT), wearable technologies, Augmented Reality (AR)/Virtual Reality (VR) & Mixed Reality (MR), BIM technologies, conversational-AI, and Robotics, visual data is interpreted and understood, and decisions in identifying and managing risks are made based on the visualised object [143, 144].

4.2.1 3D/4D visualised technologies

During the design phase, 3D and 4D visualisation technologies are used to assist the teams in identifying risks regarding preliminary construction activities. Building Information Modelling (BIM) is a 3D/4D modelling technology that is used prior to the construction planning process to create, assess, and manage design and construction information, including planning for risk management. [145, 146] in their investigations shows that projects using 3D and 4D visualisation technology enabled the designing and developing team to detect potential risks in the early stages of construction. Early risk detection leads to cost savings and minimum design errors. Consequently, the likelihood of experiencing risk-related problems during all phases of the project's life cycle is decreased. Utilising 4D visualisation technologies aids in detecting improper construction processes and procedures, establishing patterns regarding risks, and distinguishing the risk status of each aspect of construction activities. Example of such activities include potential risks during excavation [147] and risks related to onsite workspace/site set-up [148]. The ability to visualise potential risks using 3D/4D visualisation technology is an excellent real-time information and communication tool to support interaction between people and the organisation. Additionally, it contributes to knowledge acquisition and training for construction risks. As BIM improves communication and understanding across cultures, socio-economic classes, and language barriers, workers receive useful risk information in workshops and training. In addition, BIM has the capability of evaluating and updating employees' levels of comprehension in various areas of construction operations and identifying risks.

Furthermore, effective training and communication among foreign construction employees have been discovered to be a developing challenge due to the dispersed locations of most construction sites. According to [149], using 3D visualisation can assist various site workers in overcoming the difficulties presented by language and cultural barriers. Several 3D visualisation applications, like interactive and nonverbal simulation, aid in improving learning among foreign employees and facilitate efficient risk training. By employing visualisation technology, risk, and safety information, along with lessons from previous projects, may be communicated and provided to the workers regardless of site location. The components of 3D visualisation, such as general visual representation, animation, and a 3D interactive viewing environment, reduce the amount of spoken description needed to generate appropriate understanding among foreign construction workers. Additionally, the techniques of training used for foreign workers are insufficient, monotonous, and not considered helpful by them. So the adoption of visualised technologies can be helpful to minimise this issue.

4.2.2 Web-based technologies

Another method for managing and disseminating accurate information on risks is the web-based communication platform. It allows project stakeholders to upload and download data about various aspects of the construction process including the risks involved [150]. A study by [151] focused on wireless telecommunications, infrared sensors, and ultrasonics to reduce the number of fatal accident on construction sites. According to him, the system functions as a preventive measure, that works by double-checking with other preventive measures already in place to reduce risk zones. Wireless internet and video technology/camera have also proven effective in identifying risks on construction sites. It works by capturing the photographs of the work site and spot any potential risks. The cameras are mounted on the roofs of nearby structures where the contractor can use the images that were taken to document the area, identify risks, and work to mitigate those risks. In the case that an accident did occur due to an unknown cause, the contractor can quickly determine what went wrong to avoid a similar occurrence through the captured images.

4.2.3 IoT technologies

The Integration of BIM with IoT is another application for onsite monitoring and risk management. According to [152, 153]. BIM and sensor-like Ultra-Wide Band (UWB) technologies are used to track the positions of site employees, materials, equipment, and the progression of activities. Sensor data and BIM models are used to analyse risk and safety by providing real-time and post-event visualisation [153]. Structure monitoring sensors, RFID tags, and BIM models enabled the visualisation of non-functioned components for structure risk monitoring [154–156]. GPS and GIS could provide location-specific services for real-time tracking systems for materials, equipment installed on the site and personnel [157] to identify potential risks.

4.2.4 Wearable technologies

Wearable technology offers the potential to provide real-time data on the construction project and potential risk areas through a results-oriented data gathering and analysis methodology. The collection of real-time data that can be tracked especially in deliveries and movement of materials, tracking of processes to keep workers safe and alert them of imminent danger, optimise working patterns and identify higher-risk movement on site. Wearables such as smart glasses and virtual reality headsets can provide all-around viewing of the construction process, job site plan accuracy, and real-time communication during construction. An example is alert systems which include mobile alert systems for group SMS function [158], sensor alert systems for onsite equipment in equipment malfunctioning [159], equipment operator proximity risks alert system [160], equipment-worker proximity alerts [161], and audio, visual or vibration alerts at entry or exit locations to work using mobile passive radio frequency identification (RFID) [162].

4.2.5 Immersive technologies

Immersive technologies and BIM enable the provision of 3D digital content and animations as real-world projections for improved contextualisation [163, 164] significant to envisioning the construction process. They help to provide an excellent platform for immersive technologies (virtual reality and augmented reality) by data-rich 3D models for intuitive visualisation of the process [165] to easily identify potential risks as the construction process progresses.

4.2.6 AI technologies

Advancements in AI, such as conversational AI have led to the creation of virtual assistants such as Microsoft Cortana, Apple Siri, Chatbots etc., which use Automatic Speech Recognition (ASP) and Text-to-Speech (TTS) to communicate construction operations and risks information to construction personnel working on site. Evidence shows that these virtual assistants and other speech processing applications have tremendous potential when applied to the construction industry. Natural Language Processing (NLP) which is a vital area of conversational AI also contribute to risk management by extracting up-to-date information and reporting outcomes during construction through a chatbot.

However, for effective collaboration, qualification and implementation identified by [141] as the socio-technical human dimensions in risk management, all stakeholders must decide on suitable interoperable software tools [166, 167] to collaborate and share up-to-date information about the construction procedures and risks management planning on site.

5. Results and discussion

Effective risk management is vital for project success. Project participants recognise this fact and employ several risk management methods to identify and manage risks. These methods were identified in this study through extant review of the literature. The study examines risks and unsafe site practices to address the digital visualisation of risk management on construction sites. Although, construction businesses implement risk management procedures as noted in the reviewed literature; however, some procedures need to be improved and used more frequently to uncover potential opportunities and negative risks. The entire structure of managing risks revolve around risk identification. Failure to identify risks can lead to inadequacies throughout the management process, which can negatively impact the project resources and success. Evidence shows that organisations that have risk management implemented acknowledge that failure is more likely if adequate procedures are not used when identifying potential risks. The accuracy of the outcomes of the risk management process is determined by a successful risk identification procedure.

The identification methods including their benefits and challenges were examined to arrive at corresponding methods of identifying risks on construction site. Some of the identified methods in the literature are adopted singly (information gathering, risk workshop, interviewing, possible risks scenario and root cause of risks) while some require the combination of more than one method for the identification process (SWOT analysis and document review). Likewise, some of the methods are useful at managerial level while some at employee level. The corresponding methods identified

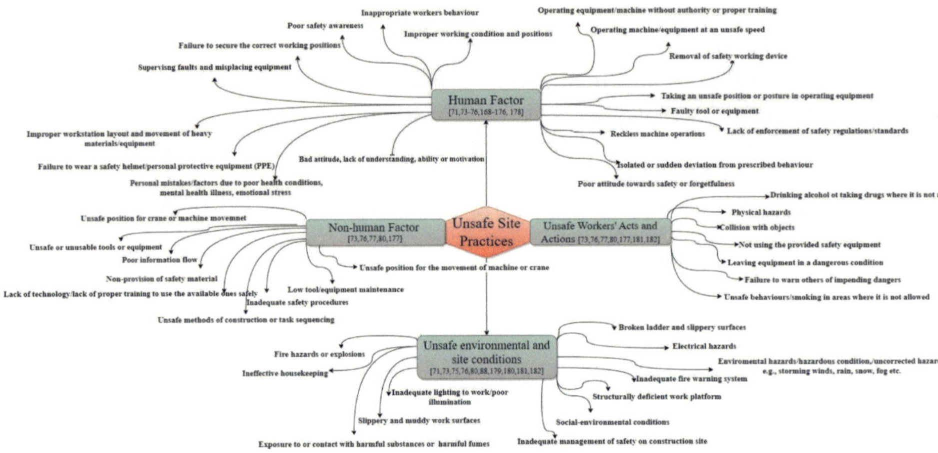

Figure 3.
Classification of unsafe site practices on construction sites [88, 168–182].

were based on the result and analysis of the reviewed literature. This includes checklists for assessing risks, portfolio of previous projects, DCPP, and identifying risks with PESTLE analysis. Sub-areas of these methods were also identified to achieve the objectives of the study. For example, identification by PESTLE analysis examines risks from six different perspectives: political, economic, sociocultural, technological, legal, and environmental. The analysis in this study examines the threat and opportunities of risks management. However, evidence shows that PESTLE analysis works well with SWOT analysis to identify internal and external risks in project management.

On the other hand, unsafe site practices were identified from four different perspectives: human factors, non-human factors, unsafe workers' acts and actions, and unsafe environmental and site conditions. **Figure 3** shows the review's findings and the resulting classification of unsafe site practices on construction sites.

The research also demonstrates the existence of numerous causes for construction risks and unsafe site practices, with human factor being the primary cause of most construction accidents. Various socio-cultural factors associated with risks management were also identified (**Table 2**). PESTLE analysis showed that risk management-related socio-cultural factors could have an impact on construction projects in the society/organisation in which they are developed. The socio-technical component of risk management on the other hand relates to people, organisations, and technology as well as their interactions to risks management. Digital technologies are designed and optimised in consideration of social, organisational structures and humans to fulfil risk-free tasks.

6. Conclusion and recommendations

The detection of potential risks and unsafe site practises on construction sites can be improved by various measures through the implementation of various visualisation technologies. It ensures that risk information is accurately and effectively provided, understood, and evenly disseminated among project managers and construction employees. Additionally, to adequately plan risk management activities and identify errors throughout the project life phases, the project manager and all participating teams must spend more time in risks management planning and identification during the planning phase. Furthermore, top management must continuously support the project's implementation, ensure appropriate site procedures, and provide training for managers and other important project players. However, it should be noted that risk management is a technique that helps to raise the likelihood of success rather than a tool that guarantees success. Risk management is, therefore, a proactive strategy rather than a reactive one.

This study provided contributions to the practical knowledge about risks identification and unsafe site practices on construction sites. It also provided valuable insight into the socio-cultural and socio-technical considerations of risk management, and their effect on construction risks. However, the major limitation of this study is that the entirety of the study was based on literature review. Alongside the review, future studies could elaborate more on the subject and develop a better risk management approach based on risks identification and unsafe site practices on construction site by employing a more comprehensive methods of data collection and analysis.

Conflict of interest

The author declares no conflict of interest.

Author details

Naimah Muhammed-Yakubu
University of the West of England, Bristol, United Kingdom

*Address all correspondence to: naimah.muhammed-yakubu@uwe.ac.uk

References

[1] Siraj NB, Fayek AR. Risk identification and common risks in construction: Literature review and content analysis. Journal of Construction Engineering and Management. 2019;**145**(9):03119004

[2] Herausgeber Project Management Institute. A Guide to the Project Management Body of Knowledge (PMBOK® Guide). Newton Square, PA: Project Management Institute; 2017

[3] Analyze I. Generic Risks Product-Specific Risks. US: Citeseer; 2014

[4] Arena M, Azzone G, Cagno E, Silvestri A, Trucco P. A model for operationalizing ERM in project-based operations through dynamic capabilities. International Journal of Energy Sector Management. 2014;**8**(2):178-197

[5] Banaitiene N, Banaitis A. Risk management in construction projects. Risk Management-current Issues and Challenges. 2012:429-448. DOI: 10.5772/51460

[6] Taroun A. Towards a better modelling and assessment of construction risk: Insights from a literature review. International Journal of Project Management. 2014;**32**(1):101-115

[7] Abazid M, Harb H. An overview of risk management in the construction projects. Academic Research International. 2018;**9**(2):73-79

[8] Rahman MM, Kumaraswamy MM. Joint risk management through transactionally efficient relational contracting. Construction Management & Economics. 2002;**20**(1):45-54

[9] Glinka S. Cross-sectional SWOT analysis of BIM and GIS integration. Geomatics and Environmental Engineering. 2022;**16**(3):157-184. DOI: 10.7494/geom.2022.16.3.157

[10] Azmy N, Zain AM. The application of technology in enhancing safety and health aspects on Malaysian construction projects. ARPN Journal of Engineering and Applied Sciences. 2016;**11**(11):7209-7213

[11] Elmsalmi M, Hachicha W, Aljuaid AM. Prioritization of the best sustainable supply chain risk management practices using a structural analysis-based approach. Sustainability. 2021;**13**(9):4608

[12] Lee I. Cybersecurity: Risk management framework and investment cost analysis. Business Horizons. 2021;**64**(5):659-671

[13] Rosa LV, Haddad AN, de Carvalho PV. Assessing risk in sustainable construction using the functional resonance analysis method (FRAM). Cognition, Technology & Work. 2015;**17**:559-573

[14] Pritchard CL, Pritchard PMP. Risk Management: Concepts and Guidance. London: Taylor & Francis Group; 2014

[15] Shayan S, Pyung Kim K, Tam VW. Critical success factor analysis for effective risk management at the execution stage of a construction project. International Journal of Construction Management. 2022;**22**(3):379-386

[16] Renuka SM, Umarani C, Kamal S. A review on critical risk factors in the life cycle of construction projects. Journal of Civil Engineering Research. 2014;**4**(2A):31-36

[17] Zavadskas EK, Turskis Z, Tamošaitiene J. Risk assessment of

construction projects. Journal of Civil Engineering and Management. 2010;**16**(1):33-46

[18] Zou PX, Zhang G, Wang J. Understanding the key risks in construction projects in China. International Journal of Project Management. 2007;**25**(6):601-614

[19] El-Sayegh SM. Project risk management practices in the UAE construction industry. International Journal of Project Organisation and Management. 2014;**6**(1-2):121-137

[20] Rostami A. Tools and techniques in risk identification: A research within SMEs in the UK construction industry. Universal Journal of Management. 2016;**4**(4):203-210

[21] Hwang BG, Zhao X, Toh LP. Risk management in small construction projects in Singapore: Status, barriers, and impact. International Journal of Project Management. 2014;**32**(1):116-124

[22] Alzoubi HM. BIM as a tool to optimize and manage project risk management. International Journal of Mechanical Engineering. 2022;**7**(1):6307-6323. ISSN 0974-5823

[23] Klemetti A. Risk Management in Construction Project Networks. Finland: Laboratory of Industrial Management Report; 2006

[24] Milosevic IN. Practical application of SWOT analysis in the management of a construction project. Leadership and Management in Engineering. 2010;**10**(2):78-86

[25] Abdul-Rahman H, Wang C, Mohamad FS. Implementation of risk management in Malaysian construction industry: Case studies. Journal of Construction Engineering. 2015;**2015**:1-6

[26] Vincent O, Chris OI, Amujiri BA, Asogwa SN, Anyadike N. The effect of SWOT analysis on project management in the Nigerian construction industry. International Journal of Physical and Social Sciences. 2015;**5**(12):51-69

[27] Purohit DP, Siddiqui NA, Nandan A, Yadav BP. Hazard identification and risk assessment in construction industry. International Journal of Applied Engineering Research. 2018;**13**(10):7639-7667

[28] Gajewska E, Ropel M. Risk Management Practices in a Construction Project–A Case Study. Swedia: Chalmers University of Technology; 2011. pp. 51-62

[29] Dziadosz A, Rejment M. Risk analysis in construction project-chosen methods. Procedia Engineering. 2015;**122**:258-265

[30] Tserng HP, Yin SY, Dzeng RJ, Wou B, Tsai MD, Chen WY. A study of ontology-based risk management framework of construction projects through project life cycle. Automation in Construction. 2009;**18**(7):994-1008

[31] Klöber-Koch J, Braunreuther S, Reinhart G. Approach for risk identification and assessment in a manufacturing system. Procedia CIRP. 2018;**72**:683-688

[32] Goh CS, Abdul-Rahman H, Abdul SZ. Applying risk management workshop for a public construction project: Case study. Journal of Construction Engineering and Management. 2013;**139**(5):572-580

[33] Thaheem MJ, De Marco A. A survey on usage and diffusion of project risk management techniques and software tools in the construction industry. In: Proceedings of World Academy of Science, Engineering and Technology.

Vol. 78. Turkey: World Academy of Science, Engineering and Technology (WASET); 2013. p. 1096

[34] Park J, Kim K, Cho YK. Framework of automated construction-safety monitoring using cloud-enabled BIM and BLE mobile tracking sensors. Journal of Construction Engineering and Management. 2017;**143**(2):05016019

[35] Yildiz AE, Dikmen I, Birgonul MT, Ercoskun K, Alten S. A knowledge-based risk mapping tool for cost estimation of international construction projects. Automation in Construction. 2014;**43**:144-155

[36] Soltanzadeh A, Mohammadfam I, Moghimbeigi A, Akbarzadeh M, Ghiasvand R. Key factors contributing to accident severity rate in construction industry in Iran: A regression modelling approach. Arhiv za higijenu rada i toksikologiju. 2016;**67**(1):47-53

[37] Eteifa SO, El-adaway IH. Using social network analysis to model the interaction between root causes of fatalities in the construction industry. Journal of Management in Engineering. 2018;**34**(1):04017045

[38] Mbachu J, Taylor S. Contractual risks in the New Zealand construction industry: Analysis and mitigation measures. International Journal of Construction Supply Chain. 2014;**4**(2):22-33

[39] Sun Y, Fang D, Wang S, Dai M, Lv X. Safety risk identification and assessment for Beijing Olympic venues construction. Journal of Management in Engineering. 2008;**24**(1):40-47

[40] Kikwasi G. Causes and effects of delays and disruptions in construction projects in Tanzania. Australasian Journal of Construction Economics and Building-Conference Series. Australia: UTS ePRESS; 2012;**1**(2):52-59

[41] Qazi A, Quigley J, Dickson A, Kirytopoulos K. Project complexity and risk management (ProCRiM): Towards modelling project complexity driven risk paths in construction projects. International Journal of Project Management. 2016;**34**(7):1183-1198

[42] Etemadinia H, Tavakolan M. Using a hybrid system dynamics and interpretive structural modelling for risk analysis of design phase of the construction projects. International Journal of Construction Management. 2021;**21**(1):93-112

[43] Royer PS. Risk management: The undiscovered dimension of project management. Project Management Journal. Eastern Cape Province, South Africa: Southeast Academic Libraries Systems (SEALS). 2000;**31**(1):6-13

[44] Bilgin G, Dikmen I, Birgonul MT, Ozorhon B. A decision support system for project portfolio management in construction companies. International Journal of Information Technology & Decision Making. 2023;**22**(02):705-735

[45] Le Roux M. Defect in the South African construction industry [doctoral dissertation]. Eastern Cape, South Africa: Southeast Academic Libraries Systems (SEALS); 2013

[46] Nourbakhsh M, Mohamad Zin R, Irizarry J, Zolfagharian S, Gheisari M. Mobile application prototype for on-site information management in construction industry. Engineering, Construction, and Architectural Management. 2012;**19**(5):474-494

[47] Niazi GA, Painting N. Significant factors causing cost overruns in the construction industry in Afghanistan. Procedia Engineering. 2017;**182**:510-517

[48] Johari S, Jha K. Determinants of workmanship: Defining quality in construction industry. In: Proceedings of the 35th Annual Conference. Leeds, UK: Leeds Beckett University; 2019. p. 761

[49] Arslan G, Kivrak S. Critical factors to company success in the construction industry. World Academy of Science, Engineering and Technology. 2008;**45**(1):43-46

[50] Al AM. Delay in payment effects on productivity of small and medium construction companies in Oman: Exploration and ranking. Asian Journal of Civil Engineering. 2021;**22**(7):1347-1359

[51] Rastogi NI, Trivedi MK. PESTLE technique–a tool to identify external risks in construction projects. International Research Journal of Engineering and Technology (IRJET). 2016;**3**(1):384-388

[52] El-Sayegh SM, Mansour MH. Risk assessment and allocation in highway construction projects in the UAE. Journal of Management in Engineering. 2015;**31**(6):04015004

[53] Chang T, Deng X, Hwang BG, Zhao X. Political risk paths in international construction projects: Case study from Chinese construction enterprises. Advances in Civil Engineering. Hindawi; 2018;**2018**:6939828. DOI: 10.1155/2018/6939828

[54] Deng X, Low SP. Exploring critical variables that affect political risk level in international construction projects: Case study from Chinese contractors. Journal of Professional Issues in Engineering Education and Practice. 2014;**140**(1):04013002

[55] Deng X, Low SP, Zhao X, Chang T. Identifying micro variables contributing to political risks in international construction projects. Engineering, Construction and Architectural Management. 2018;**25**(3):317-334

[56] Low SP, Liu JY, Ng SH, Liu X. Enterprise risk management and the performance of local contractors in Singapore. International Journal of Construction Management. 2013;**13**(2):27-41

[57] Shrestha A, Chan TK, Aibinu AA, Chen C, Martek I. Risks in PPP water projects in China: Perspective of local governments. Journal of Construction Engineering and Management. 2017;**143**(7):05017006

[58] Iyer KC, Sagheer M. Hierarchical structuring of PPP risks using interpretative structural modelling. Journal of Construction Engineering and Management. 2010;**136**(2):151-159

[59] Liu J, Zhao X, Yan P. Risk paths in international construction projects: Case study from Chinese contractors. Journal of Construction Engineering and Management. 2016;**142**(6):05016002

[60] El-Sayegh SM. Risk assessment and allocation in the UAE construction industry. International Journal of Project Management. 2008;**26**(4):431-438

[61] Zarrouk H, El Ghak T, Al Haija A, E. Financial development, Islamic finance, and economic growth: Evidence of the UAE. Journal of Islamic Accounting and Business Research. 2017;**8**(1):2-2

[62] Iranmanesh M. The impacts of environmental practice characteristics on its implementation in construction project. Procedia Environmental Sciences. 2017;**37**:549-555

[63] Swierczek FW. Culture and conflict in joint ventures in Asia. International

Journal of Project Management. 1994;**12**(1):39-47

[64] Akanni PO, Oke AE, Akpomiemie OA. Impact of environmental factors on building project performance in Delta State, Nigeria. HBRC Journal. 2015;**11**(1):91-97

[65] Lückmann P, Färber K. The impact of cultural differences on project stakeholder engagement: A review of case study research in international project management. Procedia Computer Science. 2016;**100**:85-94

[66] Han SH, Diekmann JE. Approaches for making risk-based go/no-go decision for international projects. Journal of Construction Engineering and Management. 2001;**127**(4):300-308

[67] Wang MT, Chou HY. Risk allocation and risk handling of highway projects in Taiwan. Journal of Management in Engineering. 2003;**19**(2):60-68

[68] Bahamid RA, Doh SI, Al-Sharaf MA. Risk factors affecting the construction projects in three developing countries. IOP Conference Series: Earth and Environmental Science. 2019;**244**(1):012040

[69] Befrouei MA, Taghipour M. Identification and management of risks in construction projects. American Journal of Civil Engineering. 2015;**3**(5):170-177

[70] Kassem MA, Khoiry MA, Hamzah N. Theoretical review on critical risk factors in oil and gas construction projects in Yemen. Engineering, Construction and Architectural Management. 2021;**28**(4):934-968

[71] Othman I, Majid R, Mohamad H, Shafiq N, Napiah M. Variety of accident causes in construction industry. In: MATEC Web of Conferences. Vol. 203. France: EDP Sciences; 2018. p. 02006

[72] Hon CK, Chan AP, Wong FK. An analysis for the causes of accidents of repair, maintenance, alteration, and addition works in Hong Kong. Safety Science. 2010;**48**(7):894-901

[73] Hamid AR, Razak AR, Yusof AM, Jaya RP, Zakaria R, Aminudin E, et al. Noncompliance of the occupational safety and health legislation in the Malaysian construction industry. In: IOP Conference Series: Earth and Environmental Science. Vol. 220(1). UK: IOP Publishing; 2019. p. 012043

[74] Kartam NA, Flood I, Koushki P. Construction safety in Kuwait: Issues, procedures, problems, and recommendations. Safety Science. 2000;**36**(3):163-184

[75] Abdelhamid TS. Identifying root causes of construction accidents. Journal of Construction Engineering and Management. 2000;**2000**:53

[76] Toole TM. Construction site safety roles. Journal of Construction Engineering and Management. 2002;**128**(3):203-210

[77] Tam CM, Zeng SX, Deng ZM. Identifying elements of poor construction safety management in China. Safety Science. 2004;**42**(7):569-586

[78] Lingard H, Cooke T, Gharaie E. A case study analysis of fatal incidents involving excavators in the Australian construction industry. Engineering, Construction and Architectural Management. 2013;**20**(5):488-504

[79] Yilmaz F. Monitoring and analysis of construction site accidents by using accidents analysis management system

in Turkey. Journal of Sustainable Development. 2015;**8**(2):57

[80] Kang K, Ryu H. Predicting types of occupational accidents at construction sites in Korea using random forest model. Safety Science. 2019;**120**:226-236

[81] Antoniou F, Merkouri M. Accident factors per construction type and stage: A synthesis of scientific research and professional experience. International Journal of Injury Control and Safety Promotion. 2021;**28**(4):439-453

[82] Magsi H, Torre A. Approaches to understand land use conflicts in the developing countries. The Macrotheme Review. 2013;**2**(1):119-136

[83] Mane S, Pimplikar DS. Risk assessment of BOT projects. International Journal of Computational Engineering Research. 2013;**3**(8):1-163

[84] Dinda S. Land acquisition and compensation policy for development activity. Journal of Land and Rural Studies. 2016;**4**(1):111-118

[85] Egbenta IR, Udoudoh FP. Compensation for land and building compulsorily acquired in Nigeria: A critique of the valuation technique. Property Management. 2018;**36**(4):446-460

[86] Hitzeroth M, Megerle A. Renewable energy projects: Acceptance risks and their management. Renewable and Sustainable Energy Reviews. 2013;**27**:576-584

[87] Kaklauskas A, Kaklauskas A. Passive house model for quantitative and qualitative analyses and its intelligent system. Biometric and Intelligent Decision-Making Support. 2015;**81**:87-112

[88] Van der Linden S. The social-psychological determinants of climate change risk perceptions: Towards a comprehensive model. Journal of Environmental Psychology. 2015;**41**:112-124

[89] Zouher Al-Sibaie E, Mohammed Alashwal A, Abdul-Rahman H, Kalsum ZU. Determining the relationship between conflict factors and performance of international construction projects. Engineering, Construction and Architectural Management. 2014;**21**(4):369-382

[90] Andrić JM, Wang JY, Zou PX, Zhong RY. Critical risk identification in one belt-one road highway project in Serbia. In: Proceedings of 22nd International Conference on Advancement of Construction Management and Real Estate. UK: Elsevier Journals; 2017. pp. 252-259

[91] Andrić JM, Wang J, Zou PX, Zhang J, Zhong R. Fuzzy logic–based method for risk assessment of belt and road infrastructure projects. Journal of Construction Engineering and Management. 2019;**145**(12):04019082

[92] Van den Heuvel G. The parliamentary enquiry on fraud in the Dutch construction industry collusion as concept between corruption and state-corporate crime. Crime, Law, and Social Change. 2005;**44**:133-151

[93] Flysjö L. A Review of Corruption and Organized Crime in the Construction Industry. Sweden: Halsa och samhalle, Malmo universitet; 2010

[94] Hertogh M. Crime and custom in the Dutch construction industry. Legisprudence. 2010;**4**(3):307-326

[95] Kankaanranta T, Muttilainen V. Economic crimes in the construction

industry: Case of Finland. Journal of Financial Crime. 2010;**17**(4):417-429

[96] Savona EU. Infiltration of the public construction industry by Italian organised crime. In: Situational Prevention of Organised Crimes. Abingdon, Oxfordshire: Routledge, Taylor & Francis Group; 2013. pp. 148-168

[97] Lohne J, Drevland F. Who benefit from crime in construction? A structural analysis. In: 10th Nordic Conference on Construction Economics and Organization. England: Emerald Publishing Limited; 2019. pp. 163-170

[98] Lipsky DB, Farber HS. The composition of strike activity in the construction industry. ILR Review. 1976;**29**(3):388-404

[99] Kaliba C, Muya M, Mumba K. Cost escalation and schedule delays in road construction projects in Zambia. International Journal of Project Management. 2009;**27**(5):522-531

[100] Enshassi A, Al-Najjar J, Kumaraswamy M. Delays, and cost overruns in the construction projects in the Gaza strip. Journal of Financial Management of Property and Construction. 2009;**14**(2):126-151

[101] Gall G. Industrial conflict in the engineering construction industry in Britain. Construction Management and Economics. 2012;**30**(7):535-544

[102] Nowak J. Mass strikes in the Brazilian construction sector, 2011-2014. Workers' Movements and Strikes in the Twenty-First Century. London: Rowman & Littlefield. 2018;**5**:115-132

[103] Moodley K, Preece C. Community interaction in the construction industry. Corporate Social Responsibility in the Construction Industry. Routledge; 2013. p. 98

[104] Zewdu ZT, Aregaw GT. Causes of contractor cost overrun in construction projects: The case of Ethiopian construction sector. International Journal of Business and Economics Research. 2015;**4**(4):180-191

[105] Caro E, Berntsen L, Lillie N, Wagner I. Posted migration and segregation in the European construction sector. Journal of Ethnic and Migration Studies. 2015;**41**(10):1600-1620

[106] Danaj S, Sippola M. Organizing posted workers in the construction sector. The Outsourcing Challenge: Organizing Workers Across Fragmented Production Networks. ETUI aisbl Brussels; 2015. pp. 217-235

[107] Khan MW, Ting NH, Kuang LC, Darun MR, Mehfooz U, Khamidi MF. Green procurement in construction industry: A theoretical perspective of enablers and barriers. In: MATEC Web of Conferences. Vol. 203. France: EDP Sciences; 2018. p. 02012

[108] Hughes WP. Identifying the environments of construction projects. Construction Management and Economics. 1989;7(1):29-40

[109] Said H, Berger L. Future trends of sustainability design and analysis in construction industry and academia. Practice Periodical on Structural Design and Construction. 2014;**19**(1):77-88

[110] Zhang S, Sunindijo RY, Frimpong S, Su Z. Work stressors, coping strategies, and poor mental health in the Chinese construction industry. Safety Science. 2023;**159**:106039

[111] Gorse CA. Conflict and conflict management in construction. In:

Proceedings of 19th Annual Conference of the Association of Researchers in Construction Management. UK: ARCOM; 2003. pp. 173-182

[112] Sunindijo RY, Hadikusumo BH. Emotional intelligence for managing conflicts in the sociocultural norms of the Thai construction industry. Journal of Management in Engineering. 2014;**30**(6):04014025

[113] Lee C, Won JW, Jang W, Jung W, Han SH, Kwak YH. Social conflict management framework for project viability: Case studies from Korean megaprojects. International Journal of Project Management. 2017;**35**(8):1683-1696

[114] Kakar AS, Hasan A, Jha KN, Singh A. Project cost performance factors in the war-affected and conflict-sensitive Afghan construction industry. Journal of Engineering, Design and Technology. 2022:1-21. DOI: 10.1108/JEDT-11-2021-0657 [Ahead-of-print]

[115] Vee C, Skitmore C. Professional ethics in the construction industry. Engineering, Construction and Architectural Management. 2003;**10**(2):117-127

[116] Bowen P, Akintoye A, Pearl R, Edwards PJ. Ethical behaviour in the South African construction industry. Construction Management and Economics. 2007;**25**(6):631-648

[117] Sambasivan M, Soon YW. Causes and effects of delays in Malaysian construction industry. International Journal of Project Management. 2007;**25**(5):517-526

[118] Chan AP, Yeung JF, Yu CC, Wang SQ, Ke Y. Empirical study of risk assessment and allocation of public-private partnership projects in China. Journal of Management in Engineering. 2011;**27**(3):136-148

[119] Han SH, Yun S, Kim H, Kwak YH, Park HK, Lee SH. Analysing schedule delay of mega project: Lessons learned from Korea train express. IEEE Transactions on Engineering Management. 2009;**56**(2):243-256

[120] Hartmann A, Bresnen M. The emergence of partnering in construction practice: An activity theory perspective. The Engineering Project Organization Journal. 2011;**1**(1):41-52

[121] Sebastian R, Böhms M, Van Den Helm P. BIM, and GIS for low-disturbance construction. In: Proceedings of the 13th International Conference on Construction Applications of Virtual Reality. Luxembourg: ITC SciX; 2013. pp. 30-31

[122] Muhwezi L, Acai J, Otim G. An assessment of the factors causing delays on building construction projects in Uganda. International Journal of Construction Engineering and Management. 2014;**3**(1):13-23

[123] Shan L, Ann TW, Wu Y. Strategies for risk management in urban–rural conflict: Two case studies of land acquisition in urbanising China. Habitat International. 2017;**59**:90-100

[124] Park JR, Park YK, Kim SB. Lesson learned from national policy construction project. CEO Information. 2005;**491**

[125] Patrick HA, Kumar VR. Managing workplace diversity: Issues and challenges. SAGE Open. 2012;**2**(2):2158244012444615

[126] Rau BL, Hyland MM. Corporate teamwork, and diversity statements in college recruitment brochures: Effects

on attraction 1. Journal of Applied Social Psychology. 2003;**33**(12):2465-2492

[127] Amadi B. Effects of workplace diversity practices on civil construction projects in Nigeria. PM World Journal. 2020;**IX**(1):1-36. ISSN: 2330-4480

[128] Acharya NK, Dai Lee Y, Man IH. Conflicting factors in construction projects: Korean perspective. Engineering, Construction, and Architectural Management. 2006;**13**(6):543-566

[129] Zhao X, Hwang BG, Lee HN. Identifying critical leadership styles of project managers for green building projects. International Journal of Construction Management. 2016;**16**(2):150-160

[130] Wood C. Environmental Impact Assessment: A Comparative Review. London: Routledge; 2014

[131] Glucker AN, Driessen PP, Kolhoff A, Runhaar HA. Public participation in environmental impact assessment: Why, who and how? Environmental Impact Assessment Review. 2013;**43**:104-111

[132] Miao J, Huang D, He Z. Social risk assessment and management for major construction projects in China based on fuzzy integrated analysis. Complexity. 2019;**2019**:1-7

[133] Yu T, Shen GQ, Shi Q, Lai X, Li CZ, Xu K. Managing social risks at the housing demolition stage of urban redevelopment projects: A stakeholder-oriented study using social network analysis. International Journal of Project Management. 2017;**35**(6):925-941

[134] Samantra C, Datta S, Mahapatra SS. Fuzzy based risk assessment module for metropolitan construction project: An empirical study. Engineering Applications of Artificial Intelligence. 2017;**65**:449-464

[135] You J, Chen Y, Wang W, Shi C. Uncertainty, opportunistic behaviour, and governance in construction projects: The efficacy of contracts. International Journal of Project Management. 2018;**36**(5):795-807

[136] Sha K. Understanding construction project governance: An inter-organizational perspective. International Journal of Architecture, Engineering, and Construction. 2016;**5**(2):117-127

[137] Turner JR, Keegan A. Mechanisms of governance in the project-based organization: Roles of the broker and steward. European Management Journal. 2001;**19**(3):254-267

[138] Kleindorfer PR, Saad GH. Managing disruption risks in supply chains. Production and Operations Management. 2005;**14**(1):53-68

[139] Gabriel S, Grauthoff T, Joppen R, Kühn A, Dumitrescu R. Analysing socio-technical risks in implementation of industry 4.0-use cases. Procedia CIRP. 2021;**100**:241-246

[140] Ulich E. Arbeitssysteme als soziotechnische Systeme–eine Erinnerung. Journal Psychologie des Alltagshandelns. 2013;**6**(1):4-12

[141] Hobscheidt D, Kühn A, Dumitrescu R. Development of risk-optimized implementation paths for industry 4.0 based on socio-technical pattern. Procedia CIRP. 2020;**91**:832-837

[142] Hirsch-Kreinsen H, Ittermann P, Niehaus J, editors. Digitalisierung industrieller Arbeit: die Vision Industrie 4.0 und ihre sozialen Herausforderungen. Germany: Nomos Verlag; 2018

[143] Luo H, Xiong C, Fang W, Love PE, Zhang B, Ouyang X. Convolutional neural networks: Computer vision-based workforce activity assessment in construction. Automation in Construction. 2018;**94**:282-289

[144] Xu S, Wang J, Shou W, Ngo T, Sadick AM, Wang X. Computer vision techniques in construction: A critical review. Archives of Computational Methods in Engineering. 2021;**28**:3383-3397

[145] Zhou Y, Ding LY, Chen LJ. Application of 4D visualization technology for safety management in metro construction. Automation in Construction. 2013;**34**:25-36

[146] Rwamamara R, Norberg H, Olofsson T, Lagerqvist O. Using visualization technologies for design and planning of a healthy construction workplace. Construction Innovation. 2010;**10**(3):248-266

[147] Talmaki SA, Dong S, Kamat VR. Geospatial databases and augmented reality visualization for improving safety in urban excavation operations. In: Construction Research Congress 2010: Innovation for Reshaping Construction Practice. US: ASCE; 2010. pp. 91-101

[148] Mallasi Z. Dynamic quantification and analysis of the construction workspace congestion utilising 4D visualisation. Automation in Construction. 2006;**15**(5):640-655

[149] Clevenger C, del Puerto CL. Using 3D visualization to train Hispanic construction workers. In: Proceedings of the 47th ASC Annual International Conference. Omaha, NE, USA; 2011. pp. 22-27

[150] Cheung SO, Cheung KK, Suen HC. CSHM: Web-based safety and health monitoring system for construction management. Journal of Safety Research. 2004;**35**(2):159-170

[151] Lee UK, Kim JH, Cho H, Kang KI. Development of a mobile safety monitoring system for construction sites. Automation in Construction. 2009;**18**(3):258-264

[152] Teizer J, Cheng T, Fang Y. Location tracking and data visualization technology to advance construction ironworkers' education and training in safety and productivity. Automation in construction. 2013;**35**:53-68

[153] Li H, Chan G, Huang T, Skitmore M, Tao TY, Luo E, et al. Chirp-spread-spectrum-based real time location system for construction safety management: A case study. Automation in Construction. 2015;**55**:58-65

[154] Kim YW, Han SH, Yi JS, Chang S. Supply chain cost model for prefabricated building material based on time-driven activity-based costing. Canadian Journal of Civil Engineering. 2016;**43**(4):287-293

[155] Yuan X, Anumba CJ. Cyber-physical systems for temporary structures monitoring. In: Anumba C, Roofigari-Esfahan N, editors. Cyber-Physical Systems in the Built Environment. Cham: Springer; 2020. pp. 107-138. DOI: 10.1007/978-3-030-41560-0_7

[156] Zhang Y, Bai L. Rapid structural condition assessment using radio frequency identification (RFID) based wireless strain sensor. Automation in Construction. 2015;**54**:1-1

[157] Costin AM, Teizer J, Schoner B. RFID and BIM-enabled worker location tracking to support real-time building protocol and data

visualization. Journal of Information Technology in Construction (ITcon). 2015;**20**(29):495-517

[158] Siebert U, Rothenbacher D, Daniel U, Brenner H. Demonstration of the healthy worker survivor effect in a cohort of workers in the construction industry. Occupational and Environmental Medicine. 2001;**58**(12):774-779

[159] Ram SN, Pedersen C, inventors; Qualcomm Inc, assignee. Wireless system for providing critical sensor alerts for equipment. United States patent US 8,559,937. 2013

[160] Marks ED, Teizer J. Method for testing proximity detection and alert technology for safe construction equipment operation. Construction Management and Economics. 2013;**31**(6):636-646

[161] Zhu Z, Park MW, Koch C, Soltani M, Hammad A, Davari K. Predicting movements of onsite workers and mobile equipment for enhancing construction site safety. Automation in Construction. 2016;**68**:95-101

[162] Kelm A, Laußat L, Meins-Becker A, Platz D, Khazaee MJ, Costin AM, et al. Mobile passive radio frequency identification (RFID) portal for automated and rapid control of personal protective equipment (PPE) on construction sites. Automation in Construction. 2013;**36**:38-52

[163] Goulding JS, Rahimian FP, Wang X. Virtual reality-based cloud BIM platform for integrated AEC projects. Journal of Information Technology in Construction. 2014;**19**:308-325

[164] Ahmed S, Hossain MM, Hoque MI. A brief discussion on augmented reality and virtual reality in construction industry. Journal of System and Management Sciences. 2017;7(3):1-33

[165] Wang J, Hou L, Wang Y, Wang X, Simpson I. Integrating augmented reality into building information modelling for facility management case studies. In: Building Information Modeling. Vol. 11. Australia: American Society of Civil Engineers; 2015. pp. 279-304

[166] Yan J. CFD visualization: A case study for using a building information modeling with virtual reality [doctoral dissertation]. California: University of Southern California; 2017

[167] Grilo A, Jardim-Goncalves R. Value proposition on interoperability of BIM and collaborative working environments. Automation in Construction. 2010;**19**(5):522-530

[168] Steel J, Drogemuller R, Toth B. Model interoperability in building information modelling. Software & Systems Modelling. 2012;**11**:99-109

[169] Hamid AR, Noor Azmi MR, Aminudin E, Jaya RP, Zakaria R, Zawawi AM, et al. Causes of fatal construction accidents in Malaysia. In: IOP Conference Series: Earth and Environmental Science. Vol. 220. UK: IOP Publishing; 2019. p. 012044

[170] Fass S, Yousef R, Liginlal D, Vyas P. Understanding causes of fall and struck-by incidents: What differentiates construction safety in the Arabian gulf region? Applied Ergonomics. 2017;**58**:515-526

[171] Zhou Z, Goh YM, Li Q. Overview, and analysis of safety management studies in the construction industry. Safety Science. 2015;**72**:337-350

[172] Vitharana VH, De Silva S, De Silva GH. Health Hazards, Risk,

and Safety Practices in Construction Sites—A Review Study. Vol. XLVDI, No. 03. Sri Lanka: The institute of Engineers; 2015. pp. 35-44. Available from: http://iesl.nsf.ac.lk/bitstream/handle/1/1867/Engineer-2015-48(3)_35.pdf?sequence=2

[173] Chi CF, Lin SZ, Dewi RS. Graphical fault tree analysis for fatal falls in the construction industry. Accident Analysis & Prevention. 2014;**72**:359-369

[174] Hosseinian SS, Torghabeh ZJ. Major theories of construction accident causation models: A literature review. International Journal of Advances in Engineering & Technology. 2012;**4**(2):53

[175] Liu HT, Tsai YL. A fuzzy risk assessment approach for occupational hazards in the construction industry. Safety Science. 2012;**50**(4):1067-1078

[176] Beavers JE, Moore JR, Rinehart R, Schriver WR. Crane-related fatalities in the construction industry. Journal of Construction Engineering and Management. 2006;**132**(9):901-910

[177] Folkard S, Tucker P. Shift work, safety, and productivity. Occupational Medicine. 2003;**53**(2):95-101

[178] Agwu MO, Olele HE. Fatalities in the Nigerian construction industry: A case of poor safety culture. British Journal of Economics, Management & Trade. 2014;**4**(3):431-452

[179] Abdelhamid TS, Everett JG. Identifying root causes of construction accidents. Journal of Construction Engineering and Management. 2000;**126**(1):52-60

[180] Pipitsupaphol T, Watanabe T. Identification of root causes of labour accidents in the Thai construction industry. In: Proceedings of the 4th Asia Pacific Structural Engineering and Construction Conference (APSEC 2000). Singapore: Springer Nature; 2000. pp. 13-15

[181] Seker S, Zavadskas EK. Application of fuzzy DEMATEL method for analysing occupational risks on construction sites. Sustainability. 2017;**9**(11):2083

[182] Halim NN, Jaafar MH, Kamaruddin MA, Kamaruzaman NA, Singh PJ. The causes of Malaysian construction fatalities. Journal of Sustainable Science and Management. 2020;**15**(5):236-256

Chapter 4

Managing Uncertainty in the Construction Phase of Road Projects

Rouzbeh Shabani, Olav Torp, Ole Jonny Klakegg and Agnar Johansen

Abstract

Construction projects are fraught with uncertainties. This chapter's objective is to demonstrate how uncertainty is managed in the construction phase of road projects. Three aspects of uncertainty management in the construction phase are examined: people and organization, management process, and tools and techniques. The specific importance of the construction phase is based on the fast changes and complexity of this phase. A longitudinal study of seven Norwegian road projects is performed. The projects were all large in terms of cost, each with costs of more than $100 million. Researchers evaluated how uncertainty emerged during the construction phase and how major uncertainties were appraised step by step during construction. In addition, this study contributes to a new recognition of important factors, which influences risk and opportunity management in large road projects by introducing a suggested conceptual framework.

Keywords: uncertainty management, risk, construction phase, challenges, enablers

1. Introduction

The topic of this chapter is managing uncertainty in the construction phase of road projects. Road projects are characterized by high complexity and high uncertainty, making the project objectives relatively challenging to achieve. Managers need to know challenges that threaten the foundation for uncertainty management. To improve uncertainty management, we need to understand the main enablers that can improve the uncertainty management process. This will pave the way for better performance of road construction endeavors. To start with, this chapter provides theories that cast light on the nature of uncertainty and the strategies and methodologies used for its management. This chapter aims to provide a suggested conceptual framework to manage uncertainty, thereby enhancing project management's capacity to flourish in the dynamic world of construction projects.

Improved processes and productivity of road projects contribute to reduced construction costs, lower project time, shorter travel times, and higher safety for travelers and goods. Many stakeholders believe that better roads lead to a firm's productivity and therefore yield economic benefits beyond those perceived by users [1]. The construction phase of road projects embraces different uncertainties, such as soil conditions, weather conditions, and material provision and quality. Furthermore, the choice of delivery models, contracts with contractors, and competency of project team members, as well as the quality of communication among stakeholders and many others, are influencing uncertainty factors.

Numerous studies have been undertaken on construction projects' uncertainty and risk management. Despite the extensive research on the matter, construction projects persist in experiencing both cost and time overruns, as well as underruns [2]. Complexity and ambiguity may be identified as principal factors contributing to the presence of uncertainty. Uncertainty, according to Galbraith [3], originates from a lack of information when a decision has to be taken. He explained that uncertainty hinders the ability of businesses to plan proactively and make sound decisions before project execution. Risks are uncertainties with negative effects on the project's objective. Opportunities are elements with potential for positive outcomes [4]. Uncertainty will relate to different aspects of projects, such as cost, time, quality, reputation, safety, sustainability, and others [5]. Uncertainties can affect different levels of a project: strategic, tactical, and operational levels. Handling uncertainties in a project is done through different types of uncertainty analyses, including managing uncertainty in the project and by uncertainty leadership in the project owner organization [6]. Different frameworks for uncertainty management exist in the literature [7–10]. In summary, uncertainty is managed by seeking to understand up front what we do not know for sure, analyzing its potential consequences, and actively choosing remedies to increase our control over the development.

The construction phase of the projects has not received adequate attention from researchers [11]. However, it is in the early phase and construction phases of the project life cycle that risk management practices are more extensively employed compared with the conceptual or termination phases [12]. Consequently, the analysis of the construction phase, regarding its complexity and demanding nature, becomes crucial.

Effective management of uncertainty in the construction phase poses a perpetual challenge. Understanding what kind of uncertainties matter in the construction phase seen from the owner's or the contractor's side is a challenge in most projects. Finding the risk that matters and estimating the right level of contingency reserves are critical for making the right decisions in the front end and for managing the project through execution. Small reserves could lead to overruns, and large reserves could lead to spending more money and time than necessary, which again is a waste.

In the uncertainty management process, some challenges persist. Knowing these challenges is useful to effectively address them in future projects [13]. Some challenges are related to the uncertainty analysis process, such as the inadequate composition of the uncertainty analysis group, excessive details in cost estimation methods (complexity), failure to recognize opportunities, and underestimated levels of uncertainty in different phases [4]. The aforementioned challenges pertain to the analysis of uncertainties in advance. Despite knowing these challenges, only a limited number of studies have deeply explored practical obstacles in uncertainty analysis and management and proposed effective solutions for them.

A multitude of uncertainty factors affect road projects [14]. Uncertainty management includes different components, such as human and organization, processes and tools, and techniques [8, 15]. Familiarity with the different components and the challenges in uncertainty management [6] is crucial for managers, engineers, and stakeholders. It equips them with the knowledge and skills necessary to effectively manage uncertainties and contribute to the success of road projects. For better recognition of challenges and to find solutions for overcoming them, understanding the uncertainty management components is important.

Understanding different components of uncertainty management [8] offers insight into uncertainty management practices. It will help project managers with strategies on how to manage uncertainties in the construction phase of the projects. Simultaneously, it will help to identify challenges and enablers for uncertainty management. These components are inseparable from uncertainty management practices because they are interdependent and may influence one another. Not knowing them makes the management process difficult or even impossible.

There are numerous models for uncertainty and risk management [4–7, 16, 17]. The availability of models or tools is not regarded as a significant concern. A variety of models, tools, and techniques to analyze and manage uncertainty in projects are available and working if they are used right. The inadequate consideration given to the process of managing uncertainty throughout the construction phase is of greater concern. The risk management process needs to be improved. One way of doing this is by analyzing how successful projects identify and manage opportunities and risks in the construction phase in practice. This represents a research gap addressed in this chapter.

The purpose of this chapter is to introduce a suggested conceptual framework for uncertainty management in the construction phase of road projects. Increased knowledge about uncertainty factors, uncertainty management enablers and challenges, and uncertainty management components are needed. The questions addressed to answer the purpose are as follows:

Q1: How are uncertainties in the construction phase managed?

Q2: What are the main challenges and enablers in the uncertainty management of road construction projects?

The second section will introduce the basic theories and components of uncertainty management. The third section focuses on the findings from empirical studies, which cover uncertainty management and challenges and enablers in uncertainty management. Section four discusses the findings, and the last section presents the conclusion and way forward.

2. Uncertainty management in large road projects

Uncertainty is an inherent aspect of everyday life, influencing our perception of situations as either opportunities or risks. This perception greatly impacts our understanding of projects and operations, allowing us to take necessary precautions to mitigate risks associated with cost, time, and project quality. Successful projects must carefully consider the interplay between cost, time, and quality, often referred to as the "iron triangle" [18]. Uncertainties stemming from various sources have direct consequences on project cost, time, and quality, influencing decision-making and daily choices. By studying uncertainty management, we not only gain a better understanding of its impact but also acquire valuable insights into how to effectively navigate and manage uncertainties in project environments.

Uncertainty origins and drivers could be categorized into operational, strategic, and contextual risks [19]. Different drivers will occur and vary over a project's

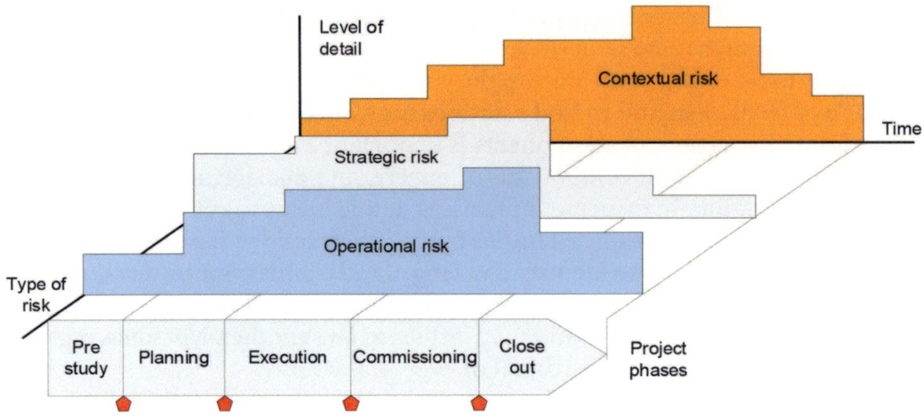

Figure 1.
Operational, strategic, and contextual risk over project phases [19].

construction phase depending on the uncertainty management strategies and the context in which the project is executed [19], as depicted in **Figure 1**. Most of the time, risk management is conducted by project managers in the pre-project and planning phases and is not followed in the construction phase. By getting more information and more analysis, the level of risk is reduced. In the execution phase, all three categories of risk are at high levels, as shown in **Figure 1**. After finishing construction, the level of risk is reduced through commissioning and closeout.

Operational uncertainties related to the project design and construction are managed by the project team. Uncertainties with strategic origins are business-level uncertainties handled by the owner or mother organization and are out of the project manager's control. Uncertainties with contextual origins are conditions external to the project, which may have an impact on its process and results, such as war and events with global effects on the whole supply chain [6] (p. 48) [8] (p. 133). During construction, certain risks may arise, which are difficult to anticipate before project initiation. Lack of competent team members, accidents, unexpected things in ground conditions, or changes in regulations could be some examples of such risks.

To effectively manage uncertainties with different origins, project managers need to have a process supported by suitable tools and expert people. Simister [20] developed a process for uncertainty management based on international standards and guidelines from professional institutions. The process is illustrated in **Figure 2**.

The process of uncertainty management includes identifying, registering, evaluating, and treating uncertainties. The three important components studied in the literature [8] for uncertainty management are as follows:

a. Human and organization

b. The process

c. Tools and techniques

In this section, we explain each of these components separately.

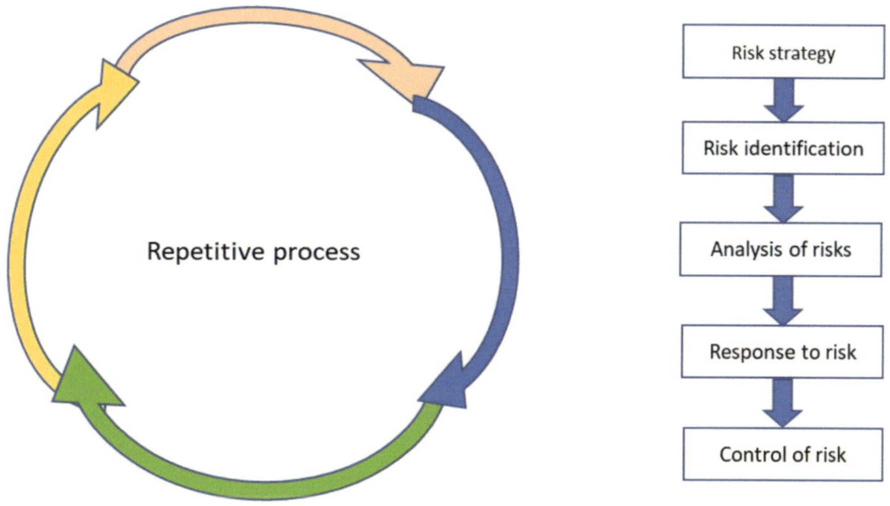

Figure 2.
Process for risk management according to Simister [20].

2.1 Human and organization in uncertainty management

The "human and organization" component incorporates a variety of professions, skills, and roles, including civil engineers, construction employees, project managers, and stakeholders. Effective uncertainty management relies on the critical elements of collaboration, coordination, and the fulfillment of different duties.

Regarding the human and organization component, risk responsibility and ownership have been emphasized as crucial factors [8, 21]. Individuals within project teams need to take ownership of the risk management processes and be accountable for their respective roles. Workforce competency is another key criterion [7, 22]. A team of individuals who possess the necessary competence and skills plays a crucial role in facilitating successful decision-making and problem-solving, particularly when confronted with unknown circumstances. The clear definition of team members' roles [6, 23, 24], effective communication [24, 25], and a supportive organizational culture [26, 27] are additional criteria that contribute to the overall effectiveness of uncertainty management.

2.2 Uncertainty management's process

The process component plays a vital role in uncertainty management, encompassing phases, such as initial planning and risk identification, risk analysis, mitigation strategies, and ongoing monitoring. Each stage of the process is interdependent, ensuring that all uncertainties are effectively addressed and mitigated.

Johansen [28] suggested a 9-step uncertainty management process for identifying, analyzing, and following up project uncertainty as illustrated in **Figure 3**.

Steps 1 and 2 are for preparing the process, Steps 3–7 are for group processes (workshops) for identifying, analyzing, and developing measures for exploiting or controlling the uncertainty, and the final steps, Steps 8–9, are for following up the uncertainty during the project life cycle. Typically, these steps are done in a risk register with a matrix consisting of opportunities and risks with a follow-up function.

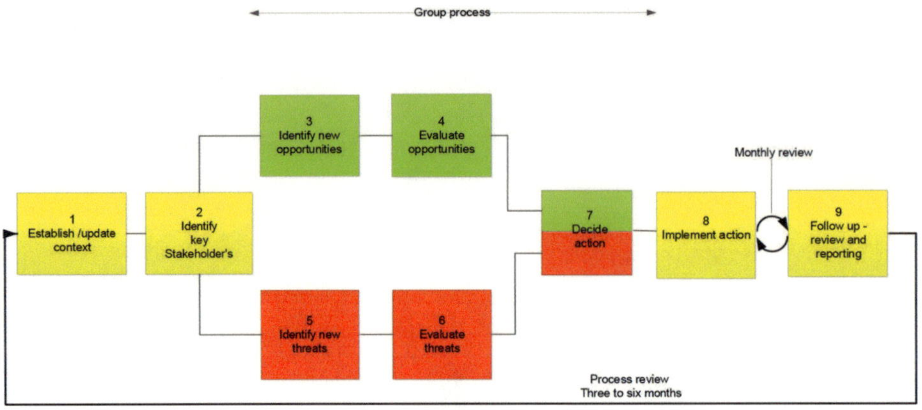

Figure 3.
Practical uncertainty management—9-step framework [28].

Formal and informal meetings have the potential to impact and facilitate the uncertainty management process. Formal meetings are pre-scheduled meetings in which people's role is defined, and project teams with various disciplines gather and conduct an uncertainty analysis session (with an external facilitator) held every month or every 6 months. These formal meetings could be project sessions in which problems and events during projects are discussed, or they could be workshops for identifying uncertainties in the projects. Informal meetings are meetings without pre-plan, such as coffee time, lunchtime meetings, or meetings in the workplace corridors [8, 29].

In the process of managing uncertainty, it is important to identify and plan for harvesting opportunities. Earlier, the focus was only on risks and threats. In recent years, the prevalent trend in risk management has shifted significantly, with a greater emphasis now placed on identifying and capitalizing on opportunities. Standardization contributes to the degree to which uncertainty management is structured and systematically followed, ensuring that all projects adhere to the same set of procedures and guidelines.

Many of these subjects that we mentioned for the uncertainty management process, such as formal or informal meetings, equal consideration for risk and opportunity, and others, originated from organizational culture. Leadership and their support play an important role. In terms of the process component, equal consideration for risk and opportunity has been identified as a crucial criterion [6]. This ensures that potential risks and opportunities are adequately assessed and addressed throughout the project lifecycle. The process should be user-friendly and simple for project teams to implement and use in the daily management of the project. Adequate documentation is essential for effective uncertainty management [7] because it facilitates knowledge sharing and provides a reference for future decision-making. Collaboration and information exchange among project stakeholders also play a vital role in the uncertainty management process [24].

2.3 Tools and techniques in uncertainty management

The "tools and techniques" component involves analytical and visualization methods and techniques for project uncertainty identification, analysis, and management.

Proper utilization of suitable tools and techniques enhances decision-making, communication, and overall project outcomes. The "tools and techniques" component requires careful consideration as well. The complexity or simplicity of the tools and techniques utilized should match the project requirements [21]. Precision is another criterion, with the need to select tools and techniques that provide accurate and reliable results [21]. The capability of visualization and documentation is also important for effective uncertainty management [8, 30].

By incorporating these criteria, stakeholders can establish a comprehensive approach to uncertainty management, ensuring the successful completion of projects while minimizing the impact of uncertainties. The cited references provide further insights into the importance of each component and serve as valuable resources for understanding and implementing effective uncertainty management practices. Additionally, the construction phase of a project is a critical stage wherein project contractors and team members are defined and resources are allocated.

The theoretical foundation presented above makes up a useful base of concepts and principles. To understand the uncertainty management practices and improve their performance in the construction phase, we turn to empirical data from real projects.

3. Empirical findings from the construction phase of road projects

The construction phase in road projects starts when the contractor and team members are in place. Project resources are decided, and a design has been conducted. At this stage, the project environment is dynamic and incorporated with different events and accidents that have the potential to influence the project, which now runs at a high speed.

Here, the empirical study reported is based on longitudinal case studies, including seven road projects in Norway. All seven projects in the study were large in terms of time and money, and they were highly complex projects. All major projects in Norway need to pass the Ministry of Finance's quality assurance (QA) to be approved by their government for financing. Documentation from this QA includes complete uncertainty analysis and is publicly available. All road projects in this study finished their construction phase, and we evaluated them in the year 2022–2023. Case studies include document studies and interviews with experts involved in the uncertainty management of these projects. The interviewees had different roles, such as project manager and project controller. The researchers studied firsthand how uncertainties were managed in the construction phase of the projects. Uncertainties and their management were structured according to three components mentioned. We present the findings in two consecutive sections.

3.1 Identifying uncertainties and managing them

Market, project organization, and project planning were the three major sources of uncertainty in seven studied projects in the construction phase. **Table 1** shows the important uncertainties in the early phase of these seven projects. The data are collected from the project's quality assurance reports or uncertainty cost analysis reports. **Table 1** contains a mixture of uncertainties, which are referred to as "factor uncertainty" that affect all or parts of the project, for example, a market that is unforeseen in relation to the level of detail, the project organization/lack of access

Project	The top 5 uncertainties
1.	• Market • Planning and engineering • Project organization • Rig of contractor for the tunnel • Water and frost protection for tunnel
2.	• Rig, contractor tunnel • Market • Project organization • Construction- Road • Construction- Road - Bridge
3.	• Market • Project organization • Contractor, rig, tunnel • Other project exercises • Unforeseen uncertainty related to the degree of detail
4.	• Market • Unforeseen uncertainty related to the degree of detail • Bracing system in bridge • Project owner cost • Bridge cable
5.	• Market • Land acquisition • Large bridge • Unforeseen uncertainty related to the degree of detail • Engineering
6.	• Project organization and losing competent personnel • Construction execution • Groundwork concerning soil/terrain/masse, measurement certificate • Owner management and framework conditions • Market
7.	• Mass balance • Blasting process • Works on long traffic roads • Geology • Environmental conditions in project (for instance: waterfall, soil, vulnerable species,...)

Table 1.
Top five uncertainties in the early phase of the projects.

to competent workforce, errors in design and poor engineering solutions, nature/weather conditions, unforeseen ground conditions, new standards and norms or laws and regulations, construction time, and geology and geotechnics.

Uncertainties in **Table 1** pertain to cost-related elements such as the rig and tunnel operation, as well as road operations. Additionally, this encompasses unpredictable circumstances characterized by events of low probability with high impact, referred to as "force majeure." In such instances, neither the owner nor the contractor can be held accountable for the occurrence of these conditions. Still, the consequences must be managed.

Market uncertainty often includes several considerations, which means that it can be complicated to assess. One way of looking at that uncertainty is to look at the difference between two different offers as a signal of how much market uncertainty there is in the offers, that is, if the lowest bidder prices the bridge at 950 million and the highest bidder prices the same bridge at 1050 million, the market uncertainty is often seen as 100 million. We believe that this initial assessment is too simple. Many different uncertainties, such as poor ground conditions, insurance requirements, and other factors can cause the latter bidder to assess differently and end up with a higher sum than the first.

Table 2 shows the comparison of uncertainties in the early phase of the seven projects with the construction phase. The uncertainties in the right column are based on interviews with the project owner and project managers from the seven case projects consisting of not only the usual uncertainties, such as the market, but also some more project-specific uncertainties, such as the replacement of bolts on project number 4. It is interesting to observe the difference among uncertainties in the early phase and the construction phase. When construction begins, the essence of uncertainties is different from the early phase. Projects face other unexpected events and uncertainties during the construction phase. In the dynamic and high pace of execution phase, project management must have strategies for tackling uncertainties that are unexpected or not identified in the early phase.

In one of the projects, the project manager emphasized that all team members should have enough competency. However, when observing the contractor's performance across two different teams involved in separate sub-projects, variations became evident. In one project, the contractor demonstrated a highly proficient understanding of the overall scope, resulting in excellent performance. Conversely, in the other project, their lack of competence hindered their ability to deliver a satisfactory outcome.

Creating a conducive working environment and establishing a robust communication platform within a project organization is vital for effective uncertainty management. "*We had good communication with the contractor, which resulted in saving money for both of us (opportunity) and it led to a win-win situation*." This situation was a bonus scheme, which was suggested to the contractor for completing part of the road. This completion of the work led to better and smoother operation and maintenance of the road, which has many benefits for the project owner too. This benefit was estimated to be approximately $1 million.

Another valuable suggestion put forth was the implementation of systematic registering of uncertainties, which involves continuous monitoring of uncertainties and aids in preventing redundant registrations leading to inefficiencies.

Furthermore, road projects should conduct comprehensive feasibility studies and scenario plans to anticipate changes in regulations, funding availability, and market conditions because of strategic uncertainties.

One of the uncertainties that could challenge projects was related to designing irrational procedures for uncertainty management. This weakness reduces our ability to identify uncertainties during the construction phase. If the uncertainties were considered in the planning phase, the result could be much better. For example, in one

Projects	Early phase	Construction phase
1.	• Market • Planning and engineering • Project organization • Rig of contractor for the tunnel • Water and frost protection -tunnel	• Injection • Little rational operation • Changes of requirements • Changes of solutions
2.	• Rig, contractor tunnel • Market • Project organization • Road Construction • Bridge	• Mass balance • Market • Weather condition • Contractor implementation competency • Frost protection
3.	• Market • Project organization • Contractor's rig for tunnel • Other project exercises • Unforeseen uncertainty related to the degree of detail	• Polluted ground • Contractors 'construction's ability • Quantity • Construction time, project organization and planning Construction time, project design
4.	• Market • Degree of detail • Bracing system in bridge • Project owner cost • Bridge cable	• Market • Progress in execution • Weather condition • Key personnel • Replacement of bolts
5.	• Market • Land acquisition • Large bridge • Unforeseen uncertainty related to the degree of the detailed engineering	• Market • Changes of routines • Supervisor and legislation • Construction in a densely populated area • Competency in its organization
6.	• Project organization and losing competent personnel. • Construction execution • Groundwork concerning soil/terrain/ masse, measurement certificate • Owner control and framework conditions • Market	• Total temporary stoppage of the project • Technical complexity in the tunnel • Construction in densely • populated areas • Ground conditions • Foundation of bridge • The technical complexity of Bridge COVID-19
7.	• Mass balance • Blasting process • Works on long traffic roads • Geology • Environmental conditions in the project (for instance: waterfall, soil, vulnerable species, and...)	• Injection • Delivery challenges • COVID-19

Table 2.
Top uncertainties in the early and the construction phase.

of the projects, the project manager said: "*There was little planning and pre-evaluation for tunneling done before the project started, and after beginning construction, unexpected things happened*".

The project manager in one of the projects said: "*Market is one of the important uncertainties, and after contracting to some extent, it is solved. Regulation and changes in them are other uncertainties which affect the project.*" Monitoring market trends regularly and maintaining solid relationships with stakeholders and regulatory bodies are helpful. Lastly, proactive engagement with stakeholders, including local communities and environmental organizations, is essential for contextual uncertainties. By incorporating stakeholders' perspectives, addressing their concerns, and incorporating their feedback into decision-making, road construction projects can reduce conflicts and improve their resilience. Using this comprehensive approach, road projects can effectively manage uncertainties.

3.2 Challenges and enablers toward uncertainty management

In this section, we presented and explored the challenges involved in uncertainty management. Solutions for these challenges can serve as enablers for uncertainty management in the context of road construction projects. The identified challenges are overemphasizing risk over opportunity management, lack of systematic approach in uncertainty management, and shortage of systematic training on uncertainty management.

3.2.1 Emphasizing risk over opportunity

In the process component, studied road projects primarily focus on risk management rather than opportunities. For example, in one of the projects, the number of risks and opportunities found in the risk register had the pattern shown in **Table 3**.

The number of risks presented in **Table 3** surpasses the number of opportunities, highlighting the significant focus on risk management in projects. **Table 4** provides an overview of the total risks and opportunities observed in the construction phase of the seven cases.

Years	Risks	Opportunities	Total
One year before construction	34	2	36
One year after construction	75	3	78
Mid construction phase	64	1	65

Table 3.
Number of identified risks and opportunities in one of the projects.

	Project 1	Project 2	Project 3	Project 4	Project 5	Project 6	Project 7
Risks	52	19	75	23	46	85	—
Opportunities	15	1	3	8	4	15	—

Table 4.
Comparison of project risk versus opportunity focus.

Data for the projects are derived from reports provided at varying intervals, such as monthly, quarterly, biannually, and yearly. Data for project 7 are not available in **Table 4** because of incomplete documentation.

In every study conducted, there was a consistent pattern of risk management taking precedence over managing opportunities. Unfortunately, this tendency persists in six projects. Addressing opportunities becomes increasingly difficult during the implementation phase because plans are decided, and in the middle of construction, identifying opportunities is difficult. The only project in which they exploited opportunities had a different form of contract and procurement strategy. The contract and procurement strategy supported opportunities in this project.

In one of the projects, the project manager mentioned: *"Due to the lack of a culture that promoted the identification of uncertainties and the pursuit of opportunities at a higher organizational level, the project ended up predominantly prioritizing risk management."*

3.2.2 To benefit from the systematic approach

Three projects from the seven in the study have a consistent strategy for managing uncertainty. They engage in a continuous process of collecting information, identifying uncertainties, implementing measures, and monitoring the uncertainties. Additionally, the project's unpredictability is reported monthly.

The case studies demonstrate varying approaches to uncertainty management. While some projects explicitly incorporate the uncertainty register into their uncertainty management practices, other projects update the register minimally after construction commences. According to the project manager's statements in the interview:

> *"I think the value of systematic risk registration is necessary for achieving better and more consistent performance. Fostering transparency and promoting effective communication among all involved parties could improve uncertainty management in projects."*

Moreover, some projects actively pursue opportunities to generate cost savings, whereas others use the register primarily to monitor the early phase identified uncertainties. Overall, each project's uncertainty register identifies more threats than opportunities. Other observed patterns in seven projects were a lack of attention to measures for identified uncertainties and not dedicating sufficient time to develop risk reduction measures. A systematic approach contributes to regularly defining and updating measures during the project's execution phase. Intriguingly, the frequency of uncertainty analysis varied significantly among projects, with some executing it monthly, others three or four times a year, and still others every 6 months. This partly demonstrates a systematic approach to uncertainty management.

Some of the projects experienced challenges in working with the risk registering tool. In some cases, some uncertainties were deleted, which affected the usage of the system for risk registration. There was no regular meeting to follow up on the uncertainty management system. One of the project managers interviewed said: *"To small extent projects logged uncertainties and measures, which they adopt for managing uncertainties. Such logs could be very helpful and could have a learning effect."*

> *The quote from the owner was: "What has been truly fascinating is that during our prolonged discussions on relationships, they frequently evolved into discussions*

> *about opportunities. This has not only helped us in terms of cost reduction but has also allowed the contractor to mitigate implementation risks and optimize resource utilization, resulting in significant cost savings and greater profitability."*

3.2.3 Systematic training on uncertainty management

The analysis of seven road construction projects revealed a lack of systematic training for those responsible for managing uncertainty. Numerous respondents mentioned initial difficulties and learning the tools and techniques through trial and error. For instance, a project controller in one of the projects said*: "We had some courses at the beginning of the career for the job position, but we did not have any* training *for uncertainty management, and we learned it during the project."* Despite confronting numerous uncertainties in their daily work, it remained unclear to what extent they utilized a systematic approach to uncertainty management. This raised the question of why businesses should invest considerable time and resources in uncertainty analysis if the project teams do not directly benefit from it. Some projects were more systematic than others in their approach to uncertainty management. If the frequency of analysis meetings increases and is supported by efficient tools, we anticipate that the outcomes will be more satisfying and in line with expectations. This will come from an increased focus on the right things and the training effect of doing it more often.

4. Discussion

This is not the first study of uncertainty in major Norwegian projects. Market, project organization, and project planning were the three major sources of uncertainty in seven studied projects in the construction phase, aligned with previous studies [14, 31]. As a reference, we note the following results: The most frequent uncertainties in the pre-project phase projects are organizational factors, market conditions, and technical conditions [31]. Project organization, market conditions, project planning and control, and technical conditions are the most frequent uncertainties [14].

Uncertainties arising from market conditions, project organization, and project planning have consistently been recognized as critical factors that can significantly impact project outcomes. Market uncertainties, such as changes in demand, price changes, competition, or economic conditions, pose challenges in predicting project success and may require adaptive strategies. Project organization uncertainties refer to factors related to the structure, roles, and dynamics of the project team, which can influence communication, decision-making, and coordination. Lastly, project planning uncertainties encompass risks associated with project scope, scheduling, resource allocation, and other planning aspects. By acknowledging the importance of these three sources of uncertainty, project managers and stakeholders can proactively identify and address potential challenges, leading to more effective risk management and improved project performance.

All the uncertainties mentioned can be categorized into operational, strategic, and contextual. To proficiently address operational, strategic, and contextual uncertainties, it would be helpful for road projects to embrace a comprehensive and integrated approach. This strategy requires forethought, effective communication, adaptability, and follow-up meetings with the participation of key people in the project team.

Comprehensive risk assessments and review plans should be conducted for operational uncertainties to mitigate potential challenges in day-to-day project activities. This includes conducting site investigations, utilizing reliable apparatus and technology, and communicating with subcontractors and suppliers openly.

Early identification of the mentioned challenges (overemphasizing risk over opportunity management, lack of systematic approach, and shortage of systematic training) in projects contributes to improving the process of uncertainty management, and thus, enhancing cost performance. By implementing effective planning and employing appropriate measures to address risks and capitalize on opportunities, project teams can work toward achieving project objectives more successfully.

In the process component, studied road projects primarily focus on risk management rather than opportunities. This pattern is in line with findings in previous studies [8, 32]. Risks are often more apparent and urgent to address because they are associated with negative outcomes, such as cost overrun and potential loss to the project. Risks are seen as more tangible and predictable, whereas opportunities require an innovative approach to identify them. Road projects also operate within tight budgets and timelines, which prioritize risk mitigation for project success. Opportunities may be perceived as adding complexity or potential delays. However, road projects must recognize and embrace opportunities for innovation and cost savings to achieve improved outcomes. Balancing risk and opportunity management can lead to enhanced stakeholder value and overall project success.

Noticing risks more than opportunities becomes particularly important when a project encounters unforeseen costs and requires cost reductions. While reduction lists and optimizations may assist in lowering expenses, they may not be sufficient to fully offset the costs incurred. In addition, because of the tight schedule and budget, there is often a low flexibility for identifying and spending time exploring opportunities in the execution phase. Failure to recognize and utilize opportunities early in projects can result in significant consequences and expensive changes later in the project's life cycle.

To improve the identification of uncertainties, project teams must recognize the significance of actively pursuing and embracing opportunities throughout the project [6]. This shift in attitude requires nurturing a culture of proactive opportunity identification and evaluation at different levels of the organization. Implementing structured processes and frameworks for opportunity management, promoting collaboration and brainstorming sessions to uncover hidden possibilities, and providing training and guidance on identifying and using opportunities are all contributing to opportunity management [8]. Improving opportunity management and effective risk management projects can maximize their success and efficiency.

The lack of a systematic approach to uncertainty management is another challenge in uncertainty management. One of the main obstacles faced in implementing a systematic approach to uncertainty management relates to the low priority given to uncertainty management processes in projects by top management, resulting in its perceived lack of importance. In some cases, there is a technical challenge wherein existing tools fail to adequately assess the impact of specific risks on project costs when uncertainties are addressed. Understanding this level of influence (uncertainty on costs) could provide valuable insights for project management in monitoring and serve as a valuable learning experience for future endeavors. Furthermore, granting access to tools for all project members enables them to proactively register risks and identify potential opportunities on a daily basis. One of the key takeaways from a particular project was the integration of uncertainty management into everyday project practices and other related activities.

As the project progresses and new uncertainties emerge, the uncertainty plans and documents should be updated regularly. This ensures that the approach remains effective and pertinent throughout the lifecycle of the project. Feedback from project team members, stakeholders, and relevant industry experts should be incorporated to refine and adapt the plans and risk registers to changing conditions.

A shortage of systematic training was obvious in seven projects. To address the lack of systematic training in uncertainty management, businesses should implement long-term training programs that cover multiple aspects of uncertainty management, foster a culture of continuous learning by encouraging ongoing professional development instead of a one-time event, and provide mentorship and support by connecting experienced professionals with less experienced team members. The focus should also be on having approximately the same goals and philosophies with uncertainty management across the projects so that there is continuity in the uncertainty work.

The level of training should be suitable for each role of the project teams to be effective. An increased focus on the transfer of experience will provide increased competence within the organization and make the project participants better equipped to find and implement the right measures to deal with uncertainties. By investing in these measures, businesses can increase the capabilities of their project teams, improve project outcomes, and boost the effectiveness of uncertainty analysis.

A comprehensive framework for uncertainty management needs three main components, as seen in **Figure 4**. This framework is a suggested conceptual framework that covers and fulfills the challenges and provides solutions (enablers) for uncertainty management.

Figure 4 is a conceptual model for achieving comprehensive uncertainty management, and it presents three components that need to work together. These suggestions depend on project size and complexity. With the increasing complexity and size of the

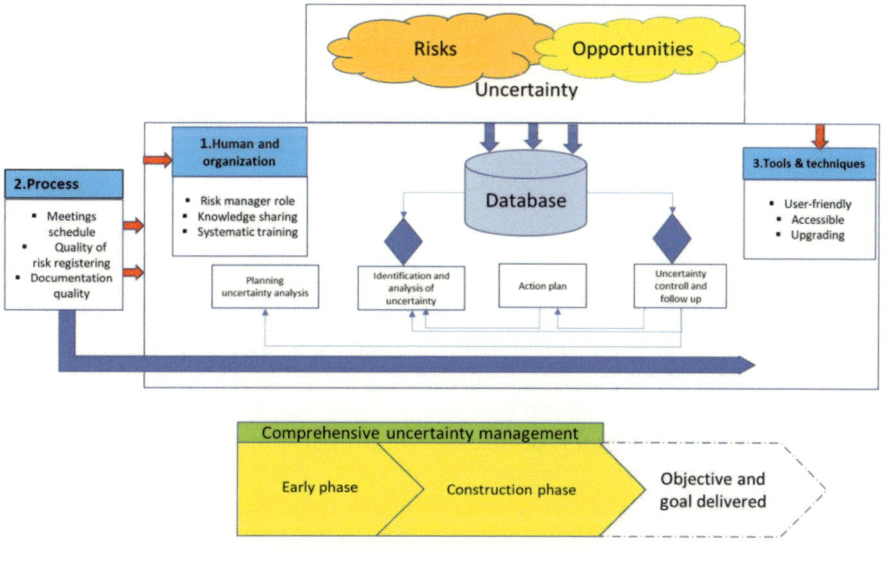

Figure 4.
The conceptual framework for comprehensive uncertainty management.

projects, the suggestions are even more important. In large and complex projects, a comprehensive framework for uncertainty management is recommended. We suggest improvements for all three necessary components of the framework.

The human and organization component covers three suggested improvement areas, such as systematic training, risk manager role, and knowledge sharing. The analysis uncovered a lack of systematic training for uncertainty management within the projects, highlighting the necessity for enhanced training initiatives and knowledge dissemination across the entire project organization. In large and complex projects, having a dedicated role as a risk manager improves coordination and communication in the uncertainty management process. This results in a more effective process. In systematic training, all the people who are involved in the uncertainty management process should have training courses regularly. When tools and processes are updated, they should have new training sessions. Besides, all new staff involved in uncertainty management should have training courses to become familiar with the main concepts of uncertainty management.

In the process components, meeting schedules, quality of risk registering, and documentation quality should be improved. Meeting schedules should include discussions on risks and opportunities. Uncertainty management meetings in projects should mirror the project size. Cost analyses and updating the risk register are typically done every 6 months in large, complex projects, and an update of the analysis of time is recommended to be done at least annually. These updates help leadership at an organizational level have a better understanding of project-level problems and obstacles during the execution process.

Tools and techniques component include a proper cost estimating tool, which is a tool for simulating time uncertainty and risk with an uncertainty/risk matrix. These tools should be an integrated part of the project reporting systems for each large complex project. The tools in use should be checked by experts and improved, if necessary, and it should be ensured that the quality of the reports is high and that the tool is user-friendly. The quality check could be conducted by project managers or personnel with enough experience in technical subjects. Conducting monthly uncertainty management activities allows the project team to consistently and systematically identify and address uncertainties. Although project team members engage in a form of uncertainty management in their daily work, formalizing this process in a systematic manner can further enhance project outcomes. The uncertainty register needs to be updated every year to remain relevant at an operational level. The risk management team needs to have a monthly focus on the top ten risks and opportunities.

5. Conclusions and the way forward

Market conditions, lack of competence in the project organization, and poor project planning are identified as the primary sources of uncertainty during the construction phase of road projects. Consistently, it has been acknowledged that these uncertainties are vital and can have a significant impact on project outcomes. To effectively manage these uncertainties, a comprehensive uncertainty management strategy could be helpful. Strategic uncertainties necessitate feasibility studies and scenario planning, whereas operational uncertainties can be addressed through comprehensive risk assessments and contingency plans. It is possible to manage contextual uncertainties such as market conditions by choosing the right contract strategy, involving stakeholders, and incorporating their perspectives.

The other interesting observation was the differences between uncertainties in the early phase and uncertainties in the construction phase. Most uncertainties in the construction phase were not expected or identified in the early phase. This shows the dynamic and changeable environment for the road projects, the focus opportunities, and the need to continue through the construction phase.

Three main challenges identified in this chapter were as follows: emphasizing risk management over opportunity management, lack of a systematic approach, and systematic training on uncertainty management.

There is strong evidence that many road projects spend most of their focus on risk management as opposed to opportunity management. Improvements can be made by measuring and changing the process, focusing on opportunity before risk. In addition, the design of contract strategies and how to share the risks and opportunities among different parties must be well thought through. Projects can achieve better performance by improving their ability to navigate uncertainty and achieve positive outcomes. Creating a harmonious synergy between a skilled workforce, appropriate tools, and a well-defined process can serve as a solution for achieving improved performance in uncertainty management.

In the future, employing quantitative approaches in uncertainty management research holds significant potential for enhancing uncertainty management practices. Through empirical studies and case studies, valuable insights into different aspects of uncertainty management can be addressed, such as identifying uncertainties, registering them, and implementing effective measures to address them. Additionally, conducting research on uncertainty management across various contract strategies and comparing their performance could yield valuable findings. Such studies can shed light on the risk management effectiveness of different contract types, providing valuable knowledge for future projects.

Acknowledgements

Thanks to The Public Roads Administration of Norway who gave us access to cases and the informants and time to answer our questions.

Conflict of interest

The authors declare no conflict of interest.

Author details

Rouzbeh Shabani*, Olav Torp, Ole Jonny Klakegg and Agnar Johansen
Department of Civil and Environmental Engineering, Norwegian University of Science and Technology, Norway

*Address all correspondence to: rouzbeh.shabani@ntnu.no

References

[1] Welde M, Tveter E. The wider local impacts of new roads: A case study of 10 projects. Transport Policy. 2022;**115**(November 2021):164-180. DOI: 10.1016/j.tranpol.2021.11.012

[2] Welde M, Odeck J. Cost escalations in the front-end of projects—Empirical evidence from Norwegian road projects. Transport Reviews. 2017;**37**(5):612-630

[3] Galbraith JR. Organization design: An information processing view. Interfaces. 1974;**4**(3):28-36

[4] Wideman RM. A Guide to Managing Project Risk and Opportunities. Vol. 1. Newton Square, PA: Project Management Institute; 1992

[5] Atkinson R, Crawford L, Ward S. Fundamental uncertainties in projects and the scope of project management. International Journal of Project Management. 2006;**24**(8):687-698

[6] Johansen A, Olsson NOE, George J, Asbjørn R. Project Risk and Opportunity Management. Routledge; 2019. pp. 67-83. DOI: 10.1201/9780429430589-6

[7] Chapman C, Ward S. Project Risk Management: Processes, Techniques and Insights. 2nd ed; 2007. Available from: https://www.amazon.com/Project-Risk-Management-Processes-Techniques/dp/0470853557

[8] Johansen A. Project Uncertainty Management: A New Approach—The "Lost Opportunities" Practical Uncertainty Management Seen from a Project Joint Perspective. Trondheim, Norway: NTNU Doctoral Dissertation; 2015;**6**:1-199

[9] PMBOK. A Guide to the Project Management Body of Knowledge (PMBOK® Guide). Pennsylvania: Project Management Institute. 5th ed. 2013

[10] Hedeman B, Seegers R. Prince2: A Pocket Guide. 1st edition. 2009. Available from: https://www.amazon.com/PRINCE2-2009-Prince2-Hedeman-Seegers/dp/B00CF6CEIS [Accessed 17 September 2023]

[11] Zheng EZH, de Carvalho MM. Managing uncertainty in projects: A review, trends and gaps. Revista de Gestão e Projetos. 2016;**7**(2):95-109

[12] Zayed T, Amer M, Pan J. Assessing risk and uncertainty inherent in Chinese highway projects using AHP. International Journal of Project Management. 2008;**26**:408-419

[13] Torp O, Klakegg OJ. Challenges in cost estimation under uncertainty—A case study of the decommissioning of Barsebäck nuclear power plant. Administrative Sciences. 2016;**6**(4):14

[14] Shabani R, Torp O, Klakegg OJ, Johansen A. Knowledge about the origins of uncertainties from the pre-project phase of road projects. Infrastructures. 2023;**8**(1):1-16

[15] Shabani R, Malvik TO, Johansen A, Torp O. Dealing with uncertainties in the design phase of road projects. International Journal of Managing Projects in Business. 2022;**16**:27-57

[16] Dikmen I, Birgonul MT, Tah JHM, et al. Web-based risk assessment tool using integrated duration—Cost influence network model. Journal of Construction

Engineering and Management. 2012;**138**(September):1023-1034

[17] Poh YP, Tah JHM. Integrated duration-cost influence network for modelling risk impacts on construction tasks. Construction Management and Economics. 2006;**24**(8):861-868

[18] Pinto JK. Project Management: Achieving Competitive Advantage. 3rd editon ed. Harlow: Pearson; 2013

[19] Rolstadås A, Johansen A. From protective to offensive project management. In: Paper Presented at the PMI Global Congress. Malta—EMEA. May 2008

[20] Simister SJ. Qualitative and quantitative risk management. In: The Wiley Guide to Managing Projects. Hoboken, New Jersey: Wiley & Sons, Inc.; 2004

[21] Klakegg OJ, Torp O, Austeng K. Good and simple—A dilemma in analytical processes? International Journal of Managing Projects in Business. 2010;**3**(3):402-421

[22] Adafin J, Rotimi JOB, Wilkinson S. An evaluation of risk factors impacting project budget performance in New Zealand. Journal of Engineering, Design and Technology. 2021;**19**(1):41-61

[23] Chapman C, Ward S. How to Manage Project Opportunity and Risk. Chichester, GB: John Wiley & Sons Ltd; 2011

[24] Osipova E. Risk Management in Construction Projects: A Comparative Study of the Different Procurement Options in Sweden. Sweden: Luleå University of Technology; 2008. pp. 1-138

[25] Aslam M, Baffoe-Twum E, Saleem F. Design changes in construction projects—Causes and impact on the cost. Civil Engineering Journal. 2019;**5**(7):1647-1655

[26] Karlsen JT. Supportive culture for efficient project uncertainty management. International Journal of Managing Projects in Business. 2011;**4**(2):240-256. DOI: 10.1108/17538371111120225

[27] Rashid A, Boussabiane H. Conceptualizing the influence of personality and cognitive traits on project managers' risk-taking behaviour. In: Wideman RM. A Guide to Managing Project Risk and Opportunities. Vol. 1. Newton Square, PA: Project Management Institute; 1992. DOI: 10.1108/IJMPB-11-2017-0138

[28] Johansen A, Halvorsen SB, Haddadic A, Langlo JA. Uncertainty management—A methodological framework beyond "the six W's.". Procedia—Social and Behavioral Sciences. 2014;**119**:566-575. DOI: 10.1016/j.sbspro.2014.03.063

[29] Krane HP, Langlo JA. Project risk management: Challenges and good practices in active project ownership. In: 24th IPMA World Congress. Istanbul, Turkey. 1-3 Nov 2010. p. 3

[30] Smith NJ, Merna T, Jobling P. Managing Risks in Construction Projects. 2nd ed. Oxford: Blackwell Publishing; 2006

[31] Torp O, Magnussen OM, Olsson N, Klakegg OJ. Kostnadsusikkerhet i store statlige investeringsprosjekter. Concept rapport No. 15. Trondheim, Norway: NTNU; 2006

[32] Hillson D. Effective Opportunity Management for Projects. New York: Marcel Dekker; 2003

Chapter 5

Risk Factors Affecting Public Infrastructure Projects

Christopher Sikhupelo and Christopher Amoah

Abstract

The delivery of public infrastructure projects in South Africa is bedevilled with many challenges leading to project delays and loss of needed public resources. This study, therefore, sought to identify the risk factors affecting project delivery and the sources of these risk factors. This study employs a qualitative research methodology. To gather the required data, open-ended interview questions were administered to the participants from the various provincial departments in the Northern Cape responsible for delivering public infrastructure construction projects. A purposive sampling technique was used to select the relevant participants to form part of the study. The data collected were analysed using qualitative content analysis. The underpinning factors for these risks affecting project execution are classified as internally and externally generated. The identified risk factors pose a significant threat to project delivery leading to delay and loss of public funds and adequate service delivery to the public. This study helps us understand the risk factors and their source for public infrastructure construction projects. The government and departments in the Northern Cape and other provinces can take measures to tackle these risk factors and alleviate their negative impact on project delivery.

Keywords: government, infrastructure, projects, public, risk factors, South Africa

1. Introduction

The public sector is the most prominent player, investor, and contributor regarding infrastructure development. The Northern Cape Provincial Government is responsible for developing and initiating public infrastructure projects to improve communities' livelihood conditions. Risks are not immune to infrastructure projects within the Northern Cape. Project Management Institute defines project risk as an uncertain situation that might constitute a threat or an opportunity for achieving the project objectives [1]. Like any other, a construction project has defined goals and lists of activities to be completed within a predetermined beginning and finish [2]. Public infrastructure projects are expected to positively contribute to the development of communities and the local economy and are faced with many negative factors, including uncertainties and risks. Some of those infrastructure projects are the construction of hospitals, roads, schools, and offices. According to the National Treasury Provincial Budgets and Expenditure Review ([3], p. 17), "Provincial Government Departments within the Northern Cape Province are mandated to deliver these infrastructure projects and related services to their sectors. Implementing agents, who usually work for provincial agencies, assist these departments; for example, the Department of Roads and Public Works is critical in planning, organising, monitoring, and supervising infrastructure projects. Infrastructure Projects directly affect the quality of life and the inclusive well-being of communities". Risk management is crucial in identifying and responding to such infrastructure projects' potential risks [3].

Siswana [4] states, "On 30th September 2019, Northern Cape Premier Dr Zamani Saul opened the long-awaited facility that started construction in 2006. Part of the facilities taking too long was because of unexpected and repeated delays." In his opening address, the Premier of the Northern Cape Province said, "What worries me is the fact that the implementation of this mental hospital, of this massive infrastructure project, reflects some major weaknesses in our capacity to implement infrastructure projects on time and budget." Infrastructure development plays a significant role in the growth of any economy. The Northern Cape Province is the largest province with a mass land, but small in the economic activity of South Africa. Over the last three years, the Northern Cape's total Provincial Infrastructure budget has decreased from R 3079 billion (Three billion and seventy-nine million rands) in 2017/18 to R 2846 billion (Two billion, eight hundred and forty-six million rands) in 2018/19. There has been a decrease in the allocation of incentives by sectors because of poor infrastructure planning and performance [5].

The causes of poor planning of infrastructure could negatively impact the performance and success of infrastructure construction within the Northern Cape, which may pose a risk. Many studies on project risk have focused on the risk management process and implementation. However, they have not yet delved into the main contributory factors to project risk leading to project failures in public infrastructure delivery. Again, most of these studies need to be more generic and focus on the provincial level to ascertain their unique risk factors and sources causing project failures [6–10]. South Africa's government spends money on public infrastructure projects to address the needs of the public. Albeit infrastructure projects within the public sector are confronted by risks that occur throughout the life cycle, the inefficient identification of risks and their sources throughout the project's life cycle would result in projects performing poorly and may lead to possible failures. Assessing risks throughout a project's life cycle can be an excellent method to make it more resilient and, as a result, more profitable for all stakeholders. This study explores the risks and causes of risks throughout the construction life cycle of infrastructure projects in the Northern Cape Province. Risks must be identified before they are controlled, managed, and mitigated. Identifying project risk aids in understanding the potential implications of risks and how to avoid them. Given the impact of risk management on a project's quality and cost, risk management goals should be an open activity throughout the project's life cycle. The subsequent sections will discuss the related literature, the methodology used to collect the relevant data, findings and the discussions and conclusion and recommendations.

2. Literature review

2.1 Project risk factors

Past studies have identified construction project risk factors; for instance, according to Al-Hazim et al. [11], cost/budget overruns are prevalent risk factors in building projects worldwide. Likewise, inconsistencies, incorrect amounts, and difficulties accessing the site were considered high-risk factors. Project risk factors include cost, scheduling, and performance [12]. Risk factors in projects must be identified and evaluated to avoid performance problems. Rezakhani [13] divided the external risk variables into unexpected and predictable uncontrolled categories. According to Abd El-Karim [14], cash flow issues, equipment shortages, late deliveries of goods, a lack of materials, and poor craftsmanship are the most significant risk-contributing variables. It is vital to note that these critical aspects may be divided into building and financing. According to Abd El-Karim [14], key risk variables should be adequately controlled to minimise the risks of failure during building projects to achieve project success. As a result, risk considerations must be evaluated to avoid causing damage resulting from decision-making errors [9]. According to Kuang [9], construction projects contain a variety of characteristics, such as unique objectives, time limits, financial needs, specific organisational and legal contractual requirements, complexity, and systematic elements. Any investment project is complex, but construction projects are considerably more due to the numerous risk factors and intricate relationships affecting the project. Therefore, incorporating risk management into building project timelines should be a crucial step [9]. The primary goals of construction project management are quality, time, cost, health and safety, and environmental sustainability. A building project's time target and cost objective are intimately and inextricably linked. South African construction is fraught with dangers. Chihuri and Pretorius [15] list some of the significant hazards connected to construction projects in South Africa: the absence of power (electricity crisis), a lack of skilled labour, and rising costs for building supplies. Similar to this, forty-four (44) risk factors for building construction were found by Kishan [8] and split into ten (10) categories, including physical, logistical, design, environmental, legal, financial, managerial, cultural, construction, and political. According to the concerns examined, design changes, poor communication, and delayed contract payments are the primary causes of project delays.

Lack of efficient planning, execution constraints, external constraints, client-induced constraints, project constraints, partner experience and a lack of project management knowledge, organisational culture and a claims redressal mechanism has been identified as major risk factors for construction projects [16]. Asumadu et al. [17] note that construction projects in wetland experience critical risk factors, including cost overruns, destruction of biological resources, water pollution, destruction of aquatic lives, lack of flood control capability and deterioration of water quality. Thus, governments should enforce effective enactment and enforcing environmental protection laws globally. Likewise, Jahan et al.'s [18] study find the rising cost of building materials, supply chain process, payment issues, planning and scheduling problems, financial difficulties, and ineffective control of manpower and equipment resources as the most critical factors affecting construction. In modelling critical risk factors in integrated construction projects, Ibrahim et al. [19] mentioned stakeholder and supply chain risks, design and capabilities risks, financing risks, and regulatory risks. In Jordan, Hiyassat et al. [20] identified the top risk factors in construction projects as client's payments delays, poorly contract forms, competition, delays in approval of permits, default by subcontractors, unclear specifications, material prices fluctuations, different construction standards, change in design, and poor design.

2.2 Public infrastructure delivery challenges in South Africa

According to academics, several difficulties stem from inadequate infrastructure development and delivery caused by various causes. Experts say South Africa's infrastructure project delivery system needs several problems. Olatunji et al. [21] state that some difficulties may arise due to wrong infrastructure delivery, including cost overruns, project delays, poor quality, inefficient fund management, and poor social benefits in service delivery. Inadequate human resources procedures, insufficient procurement methods, and a lack of leadership impede service delivery [22]. To provide public services, the government relies on various state departments [23]. According to Koops [24], effective project delivery might include additional dependent elements, such as observable performance, to meet customer expectations. According to the World Bank [25], increasing public sector performance necessitates eliminating infrastructural gaps. A rise in domestic and foreign direct investment (FDI), as well as private sector engagement via the adoption of public-private partnerships (PPPs) in public infrastructure delivery, appears to be a critical effort made by the governments [26]. According to Kudumela [27], a deficit in skills capability at the local level, lack of finance, political instability, and corruption are among the key reasons that cause problems for infrastructure delivery in underdeveloped nations. There were several obstacles to public infrastructure project delivery in South Africa. While these problems and variables apply to all types of infrastructure, public construction infrastructure needs to be immune to these issues. The Presidency [28] lists the following as enablers where communities may profit from better infrastructure project performance and where problems can be resolved via action. They listed project challenges as follows;

- Poor institutional planning;
- Slow project approval;
- Late project start;
- Poor execution quality;
- High costs and monopolistic pricing;
- Slow industry response time;
- Inadequate project controls (time, cost, quality, safety, health, and the environment);
- Tender abuse and corruption; unrealistic acceleration; unplanned and expensive design or construction redo;

- There is no political alliance and no champion;
- Allowing for rework or delays in environmental impact assessments (EIAs); and
- Delays in lead time; and slow or non-payment of contractors.

To overcome these challenges to ensure communities benefits from projects, they proposed many actions to improve the performance of infrastructure, including:

- Program to coordinate state-wide development of project-related, project management, and engineering skills;
- Align the federal, provincial, and municipal frameworks; – Align the investment strategy with the distribution of resources;
- Strong policy direction for encouraging supplier growth, localisation, and private sector engagement; Long-term support for long-term initiatives, particularly regional projects; Predictable procedure for initiating national projects, including regulatory clearances;
- Create projects with minimal life-cycle costs in mind;
- Standardised designs and delivery; recovery of full life cycle costs via user tariff or committed financing plan;
- Improving project oversight and controls in state-owned enterprises (SOEs) and the government;
- Standardised and straightforward automated reporting to monitor project performance;
- Early notification of bottlenecks; and
- A new agreement with the private sector [28].

The Infrastructure Development Bill passed Act No. 23 of 2014 on June 2, 2014, aiming to increase the government's ability to deploy infrastructure. Infrastructure development is one of the essential strategies for resolving the legacies of privilege and underdevelopment that are a part of our terrible history [29].

3. Research methodology

3.1 Description of the study area: Northern Cape Province

The Northern Cape is approximately a third of South Africa's land area and is South Africa's biggest province. It is the least populated province in South Africa, with an area of 372889 km^2 and a population of 1,193,780.00. It is bounded by Namibia, Botswana, Northwest, Free State, Eastern Cape, and Western Cape provinces. The Atlantic Ocean forms the province's western border. Kimberley is the capital city of the province. The province also has significant towns such as Upington, the centre of the karakul sheep and dried-fruit industries; Springbok, in the heart of Namaqualand spring-flower; and Kuruman and De Aar, the second most significant railway junction in South Africa. Again, Sutherland is home to the Southern African Large Telescope, the southern hemisphere's largest astronomical observatory. The Northern Cape is mineral-rich, and in Alexander Bay and Port Nolloth, alluvial diamonds can be collected from the beaches and sea.

The Sishen Mine in Kathu is South Africa's largest source of iron ore, while the copper mine at Okiep is one of the country's oldest. Springbok and Aggeneys also have copper mines. Asbestos, manganese, fluorspar, semi-precious stones, and marble are abundant in the province. The province is divided into 5 district municipalities: Frances Baard, John Taolo Gaetsewe, Namakwa, Pixley Ka Seme and ZF Mgcawu, further split into 26 local municipalities, as stated in **Figure 1**. The province was selected for this study because the province is currently experiencing infrastructure development to cope with the increasing population and economic development. Thus, identifying risk factors for infrastructure development will assist the authorities in reducing project failures and saving public resources.

3.2 Research approach

A qualitative research technique was used for this study. The qualitative research technique is appropriate since this study examines in-house peculiar risk factors from the participants with experiences and involvement in implementing and managing public infrastructure construction projects. The qualitative research approach uses

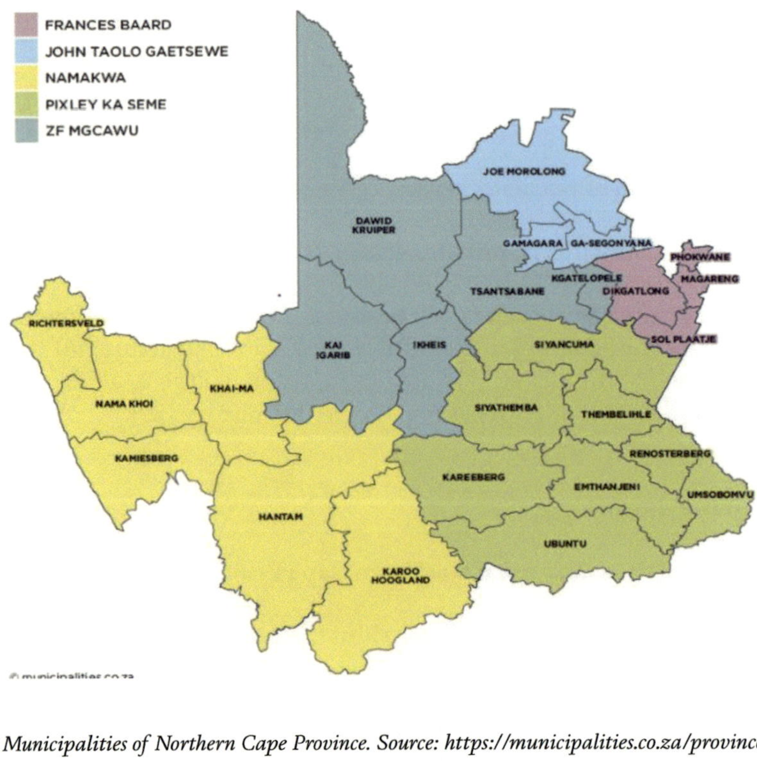

Figure 1.
Five District Municipalities of Northern Cape Province. Source: https://municipalities.co.za/provinces/view/7/northern-cape.

interviews via open-ended interview questions or guides in collecting data from the participants. Thus, participants can freely express themselves based on their experiences of the phenomena under investigation [30]. According to Creswell [30], qualitative research techniques employ an approach to investigate and comprehend the meaning that people or groups assign to social or human problems. Unlike a quantitative approach that forces participants to select from predetermined variables or responses, a qualitative research strategy may place the researcher in the world of study participants, exposing them to the richness and depth of the participants' impressions of their experiences [31]. Thus, using a qualitative research strategy allows the researcher to examine the participant's experiences regarding the risk factors confronting their projects [32].

3.3 Target population

O'Leary [33] defines population as the total unit of a particular class or group from which a sample is drawn. Bryman [34] describes a population as a collection of people or items considered for a research study. The study was conducted in the Northern Cape; however, participants were selected from various provincial government departments responsible for delivering public infrastructure projects. 28 participants from eight Departments (see **Table 1**) within the Northern Cape Provincial Government and responsible for the execution of public infrastructure projects in their respective departments were interviewed. They perform roles such as project managers, quantity surveyors, architects, planners, engineers, middle and senior management, and project stakeholders.

3.4 Sampling method

The participants for this study were chosen using the purposive sampling technique. The researcher first identified the relevant population, which comprises the entire number of elements, to extract a sample [35]. In this case, individuals from the Northern Cape provincial government departments are responsible for delivering public infrastructure projects. A reasonable population sample had to be drawn

Departments and institutions interviewed in the Northern Cape	Number of participants
Department of Roads and Public Works	10
Department of Education	5
Department of Health	5
Provincial Treasury	3
Department of Transport, Safety and Liaison	2
Department of Cooperative Governance, Human Settlement and Traditional Affairs	1
Department of Social Development	1
Department of Sports, Arts and Culture	1
Total	28

Table 1.
Participants' place of work.

after identifying the target demographic. Purposeful sampling is a technique that is frequently employed in qualitative research [36]. Purposive sampling allows researchers to find and choose people or groups with exceptional expertise or experience in the phenomenon under investigation [37]. The research focuses on the infrastructure construction projects within the Northern Cape Provincial Government. Thus, project implementers and managers from various departments in the case study province were purposively selected for the interview. According to O'Leary [28], sampling is the "process of selecting elements of a population to participate in the study." The researcher invited 32 participants for an interview, of which 28 participated. According to Leedy and Ormrod [38], a qualitative study among a homogeneous population using interviews requires a sample size of 5–25. In Camngca et al.'s [39] study, 8 participants were used, and a saturation point was realised at the 7th interviewee. Again, Smith and Amoah [40] used 16 participants for their qualitative study and reached a saturation point at the 12th interviewee. A saturation point for this study was realised at the 22nd participant; thus, 28 interviewees used for the analysis are justified. According to Mayring [41], what is essential is when saturation is reached during a qualitative interview.

3.5 Data collection methods

The research study made use of a semi-structured interview guide. The researcher conducted semi-structured interviews. The semi-structured interviews were meant to get the participants' verbal, reliable, and valid information. Letters were sent to the Heads of Department offices for the various departments to get permission and set up interview appointments. The researcher visited the targeted individual's offices to conduct the interviews at an agreed convenient time. The participants were contacted telephonically to confirm their availability before the researcher visited their offices. The researcher took notes during interviews and used a tape recorder. An interview is a data collection exercise whereby the researcher asks questions and records verbal answers from the participant [42]. Where there were ambiguities and discrepancies in the participant's responses, the researcher sought clarifications to ensure consistency in the interviewees' responses to the questions asked. This also enriched the data and information provided by the interviewees. The interview took about 30–45 minutes per participant. The saturation point (where no new information emerged) was reached at the 22nd participant. The working departments of the participants interviewed are indicated in **Table 1**. In the discussion, the participants were asked to indicate their demographic data, including their position and experience in the organisation. The participants were then asked to express their views on the risk factors that usually confront them in the project execution process, whether risk management is included in the project plan, stakeholders involved in risk assessment and risk management practices in the organisation, among others. However, this study focuses on the risk factors confronting them in the project execution.

3.6 Data analysis methods

Data analysis dissects the gathered information into components to address the research questions or achieve the research goals [43]. Data editing and coding were done by repeatedly listening to the recorded participants' responses and notes taken during the interviews. The gathered data was examined using the content analysis method. This method allows the researcher to review transcriptions of recorded data carefully. Data gathered through the interviews were analysed using Polit and Beck's [44] qualitative contents analysis processes, as indicated in **Figure 2**. The demographic features of the participants are indicated in **Table 2**.

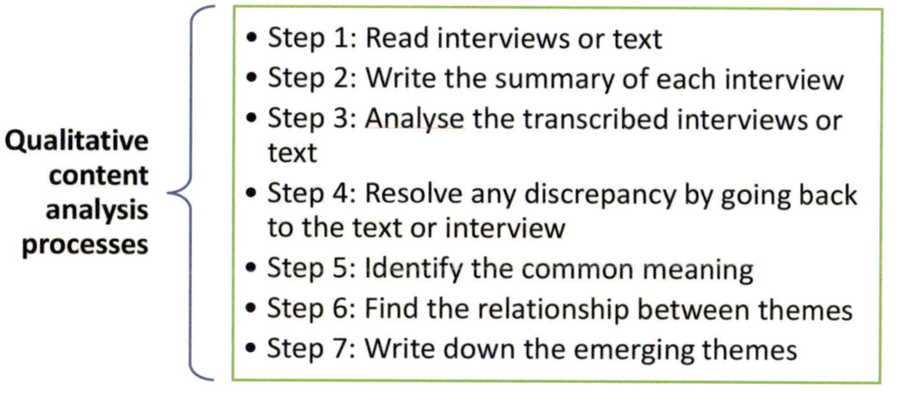

Figure 2.
Qualitative contents analysis. Source: Polit and Beck ([44], p. 542).

Demographics		Frequency	Percentage
Gender	Male	18	64%
	Female	10	36%
	Total	28	100%
Position	Project Manager	5	18%
	Director	4	14%
	Deputy Director	5	18%
	Engineer	6	20%
	Architect	3	11%
	Quantity Surveyor	1	4%
	Planner	3	11%
	Control Works Inspector	1	4%
	Total	28	100%
Experience	0–5 years	6	21%
	6–10 years	8	29%
	11–15 years	4	14%
	Over 15 years	10	36%
	Total	28	100%

Table 2.
Participants' demographic data.

Step 1: The researcher repeatedly listened to each recorded interview to understand the interviewees' statements. This was done by downloading the recorded interviews into the computer. The researcher then played the interview loudly and listened attentively to have a general overview of what the participants stated.

Step 2: The researcher transcribed the summary of each interview in a notebook and typed it into a Word document. This also assisted the researcher in further reading the participants' narrations to understand and make the statement meaningful.

Step 3: The researcher carefully analysed each transcribed data to identify any differences in the responses. This was done by carefully reading and comparing the participant responses to questions asked and notes made during the interview.

Step 4: Where differences were identified, the researcher replayed the recordings for further clarification and referred to the notes made during the interview to assist in reconciling responses.

Step 5: The researcher identified common meanings from the participants' responses and entered them into an Excel spreadsheet. This was done by using keywords or themes to define participants' responses.

Step 6: The researcher identified similar meanings from the themes identified in each interview. This was done by comparing various themes stated by each participant to see where similarities exist.

Step 7: The researcher finally merged similar themes. This was then by grouping similar themes into one to reduce the volume of themes listed. Then sub-themes were identified and grouped under the main themes.

Among the 28 participants in the study's total sample, 64% identified as male and 36% as female. Again, the position compositions of participants indicate that the majority (20%) are engineers in the departments involved in the government project execution, whilst 18% are project managers and deputy directors, respectively. Also, participants' years of experience executing public infrastructure construction projects indicate that 50% have over 10 years, while 29% have 6–10 years. This shows that participants have experiences and knowledge of the risk factors causing failure in government infrastructure projects.

4. Participants' views on the risk factors that confront project execution

The rationale behind this question was to establish if public infrastructure construction projects are confronted with risk factors. **Table 3** shows that the risk factors that usually confront projects are poor contract management, which 24 participants mentioned, representing 86%, 18 representing 64% of the participants, identified poor contractor performance as a critical risk factor confronting their projects. Again 12 participants, representing 43%, mentioned community disruptions as a significant project risk in the province.

Other project risk factors cited by the participants are political interference (cited by 39%), lack of technical skills (cited by 32%), payment delays and unforeseen events, mentioned by 29% of the participants, respectively. Likewise, 25% and 18% mentioned material availability and corruption as risk factors, respectively. It can be deduced that participants are updated with risk issues affecting their project by looking at the similar issues that different participants cited.

Question	Response	Frequency	Percentage
What are the risk factors that usually confront your projects?	Poor contract management	24	86%
	Poor performance of contractors	18	64%
	Community disruptions	12	43%
	Political Interference	11	39%
	Lack of technical skills	9	32%
	Payment delays	8	29%
	Unforeseen events	8	29%
	Material availability	7	25%
	Corruption	5	18%
	Total Participants	28	100%

Table 3.
The risk factors that usually confront projects.

5. Discussions of the findings

5.1 Risk factors affecting public infrastructure projects

The study's first objective was identifying the risk factors affecting public infrastructure projects. The risk factors identified by participants are listed in **Table 3**, with the prominent ones cited by most participants s being poor contractor performance, poor contract management and Community disruptions. The risk factors usually negatively impact a project's performance and success. Abd El-Karim [14] points out that the risk factors usually confront projects are cash flow issues, equipment shortages, late deliveries of goods, a lack of materials and poor craftsmanship contributing factors. Likewise, Kishan [8] found forty-four (44) risk factors for construction projects and were categorised into ten: managerial, cultural, construction, financial, political, logistical, design, environmental, legal and physical. This indicates that project risk factors faced by project implementers are similar. This is because participants in these stated issues, such as political interference and cashflow challenges of the municipalities executing the project, led to the delay in payment of the invoices submitted by contractors and consultants, which Kishan's findings support. Likewise, contractors have been found to mismanage the payment received from the client, leading to their inability to procure the required skilled personnel and execution of shoddy work. Heagney [45] stated that the causes for risk factors are appointed contractors graded by Construction Industry Development Board (CIDB) but do not have the skills, capacity, or technical know-how to execute the project effectively. Most of these issues emanate from political interference by people in higher authority and corrupt practices by the official during the contractor procurement process. As elaborated by P6, P9 and P12.

> **P9:** *"Competency of Contractors. Cashflow of the contractor. Unforeseen factors on site - soil conditions, the community and business forms."*

> **P12:** *"The biggest concern we have is that contractors are not performing well and this impacts negatively on the projects we have."*

P6: *"Contractors that are usually appointed intend to have a high grade level as per CIDB, but they are unable deliver the quality of work as required by their grading level."*

It has been identified by Koops [24] that effective project delivery might also include additional dependent elements, such as observable performance, to meet customer expectations. This point could be linked to what participants mentioned about contractors' poor performance, as mentioned by P12. The issue of poor contract management on the part of the project managers was also a significant issue raised by many participants. For instance, P15 and P28 suggested that, more often, it becomes challenging to understand the type of contract used to manage the project, thus hampering their ability to manage the terms and conditions of the contract properly. Likewise, most project consultants need help understanding the contract used to manage the project, hampering the ability to enforce it when the contractors flout its rules. P28, on the other hand, stated that project consultants often start projects with preliminary designs leading to scoop creep at the later stage of the projects. Oke et al. [21] mentioned some causes of bad infrastructure delivery, such as cost overruns, project delays, poor quality, poor money management and poor social benefits in service delivery. Heagney [45] believes project planning should incorporate risk assessments within predetermined project objectives. The views of P3, P15 and P28 are stated as follows:

P15*: "The management of contracts, we do not even know what are the type contracts that are utilised for example is it the JBBC or what"*

P28: *"Projects completed beyond schedule, Time and cost overruns, Implementation of projects with un-approved designs, Validity of appointment letters, Variation orders, Poor risk identification"*

P3*: "The issue of planning of projects it is a serious challenge, the planning process impacts on the entire duration of the project"*

Many participants also mentioned the issue of community disruptions of ongoing and new projects in the province. More often, groups emerge with names such as a business forum, construction mafia, etc., coming to project sites and demanding some portion of the contract amount to be allocated. Adherence to the demand by the contractor will lead to the closure of the site as they will come and damage site equipment and steel or destroy the materials on site till the project implementers meet their demands. This phenomenon is becoming rampant in the Northern Cape provinces and other areas in South Africa, posing a major risk to smooth project implementation and causing significant project delays. For instance, P10 list his project risk factors as follows:

P10*: "Contractors tender higher than on private construction projects. Interference of Construction Mafia/Construction Forums can increase tender amounts and construction time, interference with appointments of Contractors, Community interference with construction."*

P6: *"Capital availability, Price inflation of materials, Scarcity of materials, Business forum interruption, Pandemic risk and Community interruption."*

Some participants also mentioned unforeseen events such as site conditions, environmental conditions, and site challenges as contributory risk factors for their projects. More often, project consultants fail to do feasibility studies by examining the site conditions to check the suitability of the land for the intended project. Thus, issues with soil incapacity to sustain the proposed project led to delays. Likewise, the government's inability to properly acquire the land earmarked for the project from the traditional leaders led to site ownership issues when the project was about to commence. These issues were also identified by Mwamvani et al. [46], where road infrastructure projects in Malawi needed to be completed due to project site issues.

5.2 Sources of risk factors in public infrastructure projects

Based on the responses from the participants, the sources of project risk factors in Northern Cape province were categorised into 8 main project risk factors, as indicated in **Table 3**. These risk factors are categorised into eight (8) main factors: project consultants-related, contractor-related, project departments-related, political-related, community-related, materials-related, unforeseen events-related and corruption-related. The risk analysis indicates that the origins of risk factors affecting the project execution of the case study province are multi-faceted. The 8 factors were further analysed, and 48 risk factors were identified. The 48 risk factors were then grouped into internally and externally generated risk factors, as indicated in **Figure 3**. This further analysis was done by looking at the main risk factors and the respondents' statements concerning the main risk factors. This was done to further break down the main risk factors into details to assist policy implementers in quickly identifying risk factors contributing to the 8 main risk factors. Based on this, the researcher identified these 48 factors contributing to the main 8 risk factors identified as the main themes shown in **Table 3**. The risk factors were categorised into internal and external to enable project implementers to identify risk factors under and outside their control. This will assist them in implementing strategies that will best suit each risk category to help solve the problems and improve project performance. Knowing specific issues affecting you will assist in instituting specific measures suitable to address them rather than developing general strategies not peculiar to the problem. Thus, knowing the internal issues will help create internally generated steps that are workable to manage. The external issues can also be tackled using an external approach as they involve external stakeholders. Therefore, project implementers, funders, provincial government agencies, and departments must examine these risk sources and institute measures to effectively address these risk factors to deliver and achieve the project's desired outcomes.

5.2.1 Internally generated risk factors

This project's risk factors are attributable to the actions and inactions of the project initiation department. More often, projects are initiated without proper planning regarding finances earmarked for the project and coordination with other departments who are also beneficiaries of the project under construction. This often leads to cashflow challenges and delaying payments to contractors. Again, the lack of coordination among the concerned departments led to unnecessary variations to the project scope and ballooning the project cost, making it difficult to get additional funding for the changes made, resulting in a delay in project completion. Likewise, unexpected budget cut for projects was also a significant issue within the department. As a result of budgetary constraints, project funds are reduced without any notice, leading to payment problems for work

Internal Risk Factors

Project departments

- Budget cuts
- No synergy with other departments
- Poor project implementation planning
- Delayed appointment of service providers
- Scope changes
- Lack of financial planning
- Capital availability.
- Payment delays
- Late completion of projects
- Lack of skilled human resources

Project consultants/managers

- Poor contract management
- Implementing Agents
- Lack of experienced personnel
- Poor health and Safety management
- Poor Project Management
- Poor planning
- Use of unapproved design
- Poor risk identification

Corruption

- Funds mismanagement
- Fraudulent practices
- Appointment process
- Poor procurement practices

Political

- Municipality interference
- Provincial interference
- Interference in labour engagement
- Interference in contractor appointment
- Interference in consultants' appointment

Internal Risk Factors

Community

- Business forum disruption
- Construction Mafia challenges
- General community disruptions
- Lack of community involvement
- Lack of community consultation
- Strike by workers

Materials

- Scarcity of materials
- Material supply delay

Unforeseen events

- Soil conditions
- Verifying appointment letter
- Natural disaster
- Site incidence
- Environmental events
- Geological events

Contractors

- Contractors' poor performance
- Inaccurate tender price
- Skilled sub-contractors
- Incompetent contractors
- Poor cash flow management
- Shoddy workmanship
- Lack of experience
- Lack of Capacity
- Financial capacity
- Poor health and Safety practices on site
- Lack of skilled labour
- CIDB grading challenges.
- Contractor liquidation
- Lack of regulation compliance

Figure 3.
Internal and external sources of project risk factors.

done and project delays. Akanni et al. [47] mentioned clusters of project risk factors that could cause cost overruns and project delays, including financial and political. They, therefore, tasked project managers to be aware of these clusters of risk factors for effective project management and prevention of cost and time overruns.

Again, some risk factors were attributable to internal project managers of the departments in charge of the project and appointed consultants. May project managers need to gain more knowledge about project contracts, and thus, they cannot effectively enforce the contract's terms and conditions during project execution, resulting in poor project management. This phenomenon usually leads to poor contractor practices that affect work quality without punitive action against contractors. Likewise, project managers usually start the project with unfinished designs, perhaps due to pressure from their superiors, leading to scope changes and additional costs. It was identified that a lack of risk management culture within a projects department is also a threat. There are no risk management guidelines that project managers must follow to ascertain possible risk events that could affect the project adversely. Thus, challenges that could have been easily identified through effective risk management exercises emanate during project execution and delay the project duration. In the view of Adeleke [48], there is a substantial correlation between following laid down rules and regulations and construction risk management. Therefore, how project managers follow guidelines concerning effective project management has implications for the project outcomes.

Corruption-related project risk factors were also categorised under internal risk factors. It was identified that department procurement managers engage in corrupt activities by appointing unqualified contractors and inflating contract prices. These actions led to poor workmanship among contractors appointed to execute projects, causing the department further expenditure. Political interference in the contract award also contributes to risk for the case study province. The interference from political elites in the contractor, consultants, and labour engagement often results in the appointment of cronies who need more technical skills and knowledge to execute the project, causing project failures in most cases. Mansfield et al. [49] state that political elites and government can invoke their powers to influence the project outcome; thus, political leadership actions are crucial in national development. Likewise, Thomas and Martin [50] suggest that projects are subject to numerous influences, including political ones; therefore, construction managers should consider the political aspect that can derail the project's success.

5.2.2 Externally generated risk factors

These risk factors are outside the internal actions of the project departments in the Northern Cape. These risks include community actions, appointed contractors' actions, materials used for the construction and unforeseen events during the project execution. Communities where projects are executed often disrupt implementation, citing department and project managers' lack of consultation and involvement. This leads to project stoppages, resulting in project delays. In some instances, groups from the province would come to the project site demanding a portion of the contract amount to be allocated to them before the project could continue. This leads to disruptions and further delays in the project implementation. The issue of groups within a community has been identified in Nigeria. A study by Engobo [51] identified that militants often kidnap foreign construction workers, demanding a ransom before releasing them. Again, unemployed community youth demand illegal fees, usually known as "settlement", from contractors. All these practices by project community groups delay project execution by weeks and months as construction activities must be halted until the issues are resolved.

Again, it was identified that workers from the community usually demand other incentives from the government and contractors that are not part of the contract. Failure to adhere to their demands results in strike action, delaying the project.

Another external factor identified is the construction materials issues. In some instances, materials take much work to come by within the communities where the project is taking place, forcing the contractor to import materials from the capital city and other provinces. This situation also usually delays the construction process. Again, materials suppliers often need to deliver the ordered materials on time, causing financial challenges to the contractor. According to Almeida et al. [52], purchase obligations are part of contracts with suppliers and are used in construction procurements. Thus, the procurement of materials needed for the project must be a priority of the procurement agent to prevent delivery delays.

Contractor-related risk factors were very profound from the narratives of the participants. These risk factors emanated from the contractors' actions. Contractors were found to submit inaccurate tender prices for projects often and needed help to execute the project successfully upon appointment. As a result of low tender prices, contractors engaged unskilled workers to maximise their profit, leading to poor work quality. The inability of contractors to secure sufficient funding for the project leads to the liquidation of these firms when payment delays for work done. This was attributed to the malpractices in the procurements process where the financial capacities of contractors are overlooked, thereby appointment unsuitable contractors for project execution.

Likewise, because of poor tender pricing, contractors were found to be wanting in compliance with occupational health and safety regulations at the project site. This issue arises because of their inability to procure the necessary safety gadgets for their workers due to the lack of effective pricing of tender documents. Odeh and Battaineh [53] suggest that it is an arduous task for project managers and contractors to predict the financial viability of a project within an unstable economic environment. Thus, Oladapo and Olotuah [54] advise that an accurate estimate for local and international projects is necessary for successful project delivery. Unforeseen events were also attributed to external factors. Projects are often halted due to land acquisition challenges between the government and the traditional leaders, leading to delays. Likewise, unforeseen site events such as soil conditions and environmental impact assessment issues delay project implementation and, at times, increase the project budget. The issue of environmental challenges being a threat to construction was adduced by Hughes (1989), who advised project managers to be conscious of tackling environmental challenges before the project execution.

6. Conclusion and recommendation

Infrastructure development is a critical area for any country's development. Thus, governments allocate significant portions of generated revenue to infrastructure development, of which the South African government is not an exception. Since government sources of income are woefully inadequate, measures must be put in place to use the resources judiciously. However, government resources are often used with less care leading to the wastage of meagre resources, especially in public project execution. The study aimed to identify the risk factors affecting the Northern Cape Province's public infrastructure projects. Through the analysis of the pragmatic information gathered from the participant's responses, issues such as poor contractor performance; poor contract management; political interference; community disruptions; material availability; and lastly, the lack of technical skills, corruption and unforeseen events were identified as prominent project risk factors. Further data analysis indicated that participants stated forty-eight (48) risk factors originating from project consultants, contractors, project departments, politics, community, materials, unforeseen circumstances, and corruption. This will give insight to project implementers to be aware of the project risk sources and institute measures to tackle these sources. The study recommends reviewing the plan implementation process to make project risk management practices mandatory before the project is commenced to identify possible risk factors that may disrupt the project objectives. Risk management guidelines must be developed for public infrastructure construction projects in all government departments. The government must ensure corrupt practices are eliminated in the procurement processes by enforcing the procurement regulations strictly to avoid unwarranted interference, which usually leads to the appointment of unqualified project executors, leading to project failures. The limitation of the study is that data were collected from a single province in South Africa; however, the findings may be similar in order of provinces. Future studies should investigate these risk factors' effects on the province's project performance.

Funding

The researchers received no external funding.

Conflicts of interest

There is no conflict of interest.

Institutional review board statement

Not applicable.

Informed consent statement

The University of the Free State's ethics committee provided the approval letter to collect the data, including the participant's consent statement.

Data availability statement

The data for this study has been included in the article. However, the raw data will be made available upon request.

Author details

Christopher Sikhupelo and Christopher Amoah*
Department of Quantity Surveying and Construction Management, University of the Free State, Bloemfontein, South Africa

*Address all correspondence to: amoahc@ufs.ac.za

References

[1] Project Management Institute (PMI). A Guide to the Project Management Body of Knowledge (PMBOK Guide). 6th ed. Newtown Square, Pennsylvania: Project Management Institute; 2017

[2] Kerzner H. Project Management. 12th ed. Hoboken, NJ: John Wiley & Sons; 2017

[3] Yornu IK, Ackah D. Examining project risk management challenges in Ghana. Project Management & Scientific Journal. 2019;**1**(3):27-39

[4] Siswana B. Leadership and Governance in the South African Public Service: An Overview of the Public Finance Management System. Pretoria: University of Pretoria; 2019

[5] National Treasury. Provincial Budgets and Expenditure Review Supplementary Book. 2021. Available from: https://www.treasury.gov.za/publications/igfr/2021/prov/2021%20PBER%20Supplementary%20book.pdf [Accessed on December 8, 2022]

[6] Ramutsheli MP, Janse van Rensburg JO. The root causes for local government's failure to achieve objectives. Southern African Journal of Accountability and Auditing Research. 2015;**17**(2):107-118. DOI: 10520/EJC181710

[7] Joel C. Risk Management Best Practices in the Department of Trade and Industry. Johannesburg: University of Johannesburg; 2016

[8] Kishan P, Bhavsar J, Bhatt R. A study of risk factors affecting building construction projects. International Journal of Engineering Research & Technology. 2014;**3**(12):831-835

[9] Kuang Z. Risk Management in Construction Projects: Application of Risk Management in Construction Period. Horsens Campus, Denmark: Bachelor of Architectural Technology and Construction Management. Via University College; 2011

[10] Tankiso M. A cross sectoral comparison of risk management practices in selected South African organizations. Problems and Perspectives in Management. 2016;**14**(3-1):239-245. DOI: 10.21511/ppm.14(3-1).2016.10

[11] Al-Hazim N, Salem ZA, Ahmad H. Delay and cost overrun in infrastructure projects in Jordan. Procedia Engineering. 2017;**182**:18-24. DOI: 10.1016/j.proeng.2017.03.105

[12] Jaber AZ. Assessment risk in construction projects in Iraq using COPRAS-SWARA combined method. Journal of Southwest Jiaotong University. 2019;**54**(4):1-17

[13] Rezakhani P. Classifying key risk factors in construction projects. Buletinul Institutului Politehnic din lasi. Sectia Constructii, Arhitectura. 2012;**58**(2):27

[14] Abd El-Karim MSBA, El Nawawy OAM, Abdel-Alim AM. Identification and assessment of risk factors affecting construction projects. Journal of Construction Engineering. 2012;**20**(15):1-6. DOI: 10.1016/j.hbrcj.2015.05.001

[15] Chihuri S, Pretorius L. Managing risk for success in a south African engineering and construction project environment. South African Journal of Industrial Engineering. 2010;**21**(2):63-77

[16] Deep S, Bhoola V, Verma S, Ranasinghe U. Identifying the risk factors in real estate construction projects: An analytical study to propose a control structure for decision-making. Journal of Financial Management of Property and Construction. 2022;**27**(2):220-238. DOI: 10.1108/JFMPC-03-2020-0018

[17] Asumadua G, Quaigrain O-M, et al. Analysis of risks factors associated with construction projects in urban wetlands ecosystem. International Journal of Sustainable Development & World Ecology. 2023;**30**(2):198-210. DOI: 10.1080/13504509.2022.2130465

[18] Jahan S, Khan KIA, Thaheem MJ, Ullah F, Alqurashi M, Alsulami BT. Modeling profitability-influencing risk factors for construction projects: A system dynamics approach. Buildings. 2022;**12**(701):1-22. DOI: 10.3390/buildings12060701

[19] Ibrahim YW, Shen GQ, Osei-Kyei R, Agyeman-Yeboah S. Modelling the critical risk factors for modular integrated construction projects. International Journal of Construction Management. 2022;**22**(11):2013-2026. DOI: 10.1080/15623599.2020.1763049

[20] Hiyassat MA, Alkasagi F, El-Mashaleh M, Sweis GJ. Risk allocation in public construction projects: The case of Jordan. International Journal of Construction Management. 2022;**22**(8):1478-1488. DOI: 10.1080/15623599.2020.1728605

[21] Olatunji SO, Oke AE, Omoregie D, Seidu A, Adeyemi S. Effects of construction project performance on economic development of Nigeria. Journal of Economics and Sustainable Development. 2016;**7**(12):142-149

[22] van Rensburg JO. Internal Audit Capability: A Public Sector Case Study. Pretoria: University of Pretoria; 2014

[23] Construction Industry Development Board (CIDB). Provincial Budgets and Expenditure Review, National Treasury. 2016

[24] Koops L. Creating Public Value: Optimising Cooperation between Public and Private Partners in Infrastructure Projects. Netherlands: Hilversam; 2017. pp. 154-483

[25] World Bank. Beyond the Gap: How Countries Can A-Ord the Infrastructure they Need while Protecting the Planet. Washington DC: World Bank; 2018

[26] Ruiters C, Matji MP. Public-private partnership conceptual framework and models for the funding and financing of water services infrastructure in municipalities from selected provinces in South Africa. Water SA. 2016;**42**:291-305

[27] Kudumela FP. Risks and Mitigations Associated with Infrastructure Development Projects in South Africa. Johannesburg, South Africa: The University of Johannesburg; 2015

[28] The Presidency. Presidency on Infrastructure Commission Council to fast-track projects valued at R340 billion. 2020. Available from: https://www.gov.za/speeches/presidency-infrastructure-commission-council-fast-track-projects-valued-r340-billion-30-jul [Accessed on December 8, 2022]

[29] Patel L. Social Welfare and Social Development. 2nd ed. Cape Town: Oxford University Press; 2015

[30] Creswell JW. Planning, Conducting, and Evaluating Quantitative and Qualitative Research. 4th ed. Boston: Pearson; 2013

[31] Taylor S, Bogdan R. Introduction to Qualitative Research Methods: The

Search for Meanings. New York: John Wiley; 1984

[32] Merriam SB. Case Study Research in Education: A Qualitative Approach. Hoboken, New Jersey, United States: Jossey-Bass; 1988

[33] Zina O. The essential guide to doing your research project. 4th edition, London: Sage, Sage Publication, 2021

[34] Bryman A. Social Research Methods. London: Oxford University Press; 2015. pp. 40-128

[35] Leavy P. Research Design: Quantitative, Qualitative, Mixed Methods, Arts-Based, and Community-Based Participatory Research Approaches. 1st ed. New York: The Guilford Press; 2017

[36] Patton MQ. Qualitative Research and Evaluation Methods. 3rd ed. Thousand Oaks, CA: Sage Publications; 2002

[37] Creswell JW, Klassen AN, Clark VLP, Smith KC. Best practices for mixed methods research in the health sciences. Office of the Behavioural and Social Sciences Research. 2011;**2011**:1-9

[38] Leedy PD, Ormrod JE. Practical Research: Planning and Design. 11th ed. Edinburgh: Pearson; 2015

[39] Camngca VP, Amoah C, Ayesu-Koranteng E. Underutilisation of information communication and technology in the public sector construction project's implementation. Journal of Facilities Management. 2022. DOI: 10.1108/JFM-10-2021-0128

[40] Smith J, Amoah C. Barriers to the green retrofitting of existing residential buildings. Journal of Facilities Management. 2022. DOI: 10.1108/JFM-12-2021-0155

[41] Mayring P. On generalisation in qualitatively oriented research, forum qualitative Sozialforschung/forum: Qualitative. Social Research. 2007;**8**(3):1-10. DOI: 10.17169/fqs-8.3.291

[42] Bhattacherjee A. Social Science Research: Principles, Methods, and Practices. 2nd ed. Florida: Textbooks Collection; 2012

[43] Babin BJ, Zikmund WG. Essentials of Marketing Research. 6th ed. Boston, USA: Cengage Learning; 2016

[44] Polit DF, Beck CT. Nursing Research: Generating and Assessing Evidence for Nursing Practice. 10th ed. International ed. Philadelphia: Wolters Kluwer Health/Lippincott Williams & Wilkins; 2017

[45] Heagney J. Fundamentals of Project Management. 4th ed. New York: Amacom; 2012

[46] Mwamvani HDJ, Amoah C, Ayesu-Koranteng E. Causes of road projects' delays: A case of Blantyre. Malawi Built Environment Project and Asset Management. 2022;**12**(2):293-308. DOI: 10.1108/BEPAM-09-2021-0113

[47] Akanni PO, Oke AE, Akpomiemie OA. Impact of environmental factors on building project performance in Delta State, Nigeria. HBRC Journal. 2014;**11**(1):91-97. DOI: 10.1016/j.hbrcj.2014.02.010

[48] Adeleke A, Bahaudin A, Kamaruddeen A, Bamgbade J, Salimon MG, Khan MWA, et al. The influence of organisational external factors on construction risk management among Nigerian construction companies. Safety and Health at Work. 2018;**9**(1):115-124. DOI: 10.1016/j.shaw.2017.05.004

[49] Mansfield NR, Ugwu O, Doran T. Causes of delay and cost overruns in Nigerian construction projects. International Journal Project Management. 1994;**12**(4):254-260. DOI: 10.1016/0263-7863(94)90050-7

[50] Thomas EU, Martin L. Essentials of Construction Project Management. Australia: University of New South Wales Press Ltd.; 2004

[51] Engobo E. Social responsibility in practice in the oil producing Niger Delta: Assessing corporations and government's actions. Journal of Sustainable Development in Africa. 2009;**11**(2):113-115

[52] Almeida H, Hankins KW, Williams R. Risk management with supply contracts. The Review of Financial Studies. 2017;**30**(12):4179-4215. DOI: 10.1093/rfs/hhx051

[53] Odeh AM, Battaineh HT. Causes of construction delay: Traditional contracts. International Journal of Project Management. 2002;**20**(1):67-73. DOI: 10.1016/S0263-7863(00)00037-5

[54] Oladapo RA, Olotuah AO. Appropriate real estate laws and policies for sustainable development in Nigeria. Structural Survey. 2007;**25**(3/4):330-338. DOI: 10.1108/02630800710772890

Chapter 6

Approaches to Improving Occupational Health and Safety of the Nigerian Construction Industry

Kamoli Adetunji, Mohd Saidin Misnan, Mohamad Zahierruden Ismail, Farah Nadiah Abdul Rahim and Zulkiflee Abdul-Samad

Abstract

The alarming figures of occupational accidents in the Nigerian construction industry, which accounted for 39.24% of the fatal accidents between 2014 and 2014, called for the subject matter. Therefore, occupational accidents mentioned earlier prompted this study on the way forward toward the approaches to improving occupational health and safety as well as the prevention of occupational accidents in the Nigerian construction industry. The quantitative method was adopted for this study. The result-revealed approaches to improving occupational health and safety are as follows: the Establishment of the Nigerian Construction Industry Development Board (NCIDB), technical assistance and collaboration among construction professionals, skill development in the management and communication of occupational health and safety, awareness-raising and advocacy on occupational health and safety, Use of International Labour Organization (ILO) mechanism on occupational health and safety, international collaboration with other professional bodies on health and safety, proper monitoring and recording of all injuries, and adequate allocation of resources (human, financial, and technology) on Occupational Health and Safety. The study called for better participation of all stakeholders in the construction industry toward improving occupational health and safety in the workplace and ensuring necessary measures in the prevention of occupational accidents.

Keywords: accidents, construction, hazard, health, occupation, safety

1. Introduction

Over the years, the focus on occupational health and safety matters has substantially increased in many nations, particularly in developed countries. Such improvement could be ascribed to numerous factors such as new methods of technical management and standards put in place by these countries. In developing countries, no such adequate provisions have been taken to moderate accidents on construction sites [1]. Nevertheless, all sectors of the economy are connected with potential exposure to risks and hazards of injuries and illnesses [2]. Health and safety appear so easy to pronounce—just make sure individuals do not get hurt, in practice, it is difficult to accomplish health and safety in an organization. Improving occupational health and safety is a major matter across the construction industry globally [3]. Occupational accidents are sources of concern to employees, organizations, and regulatory agencies because the relationship between accidents and injury or death can be easily recognized. It is principally for this purpose that many discussions on safety and health issues are focused. The construction industry is a significant section of the economy in many nations but it is mostly labeled to be the most hazardous. Fatality and death frequency in the construction sector globally may show intrinsically poor health and safety management norms. According to Rowlinson [4], even with decades of research that has been trying to solve the problems of its failure, construction has been the poorest in terms of safety performance globally. The industry has been characterized by various issues around its division, resulting in conflicts, communication problems, and a lack of collaboration among the stakeholders. Among parts of the problems faced by the industry is from the clients with the absence of long-time view of controlling its properties and worsened by the bad organization on site and lack of appropriate complete design of temporary and permanent works. These problems affecting the sector are not only limited to safety but also include sustainability, productivity, and quality. The sector considers itself to be distinct and the issues faced by it are seen to be wicked in nature.

At the global level, the construction sector accounts for 30% of the entire fatality [5]. The construction sector is responsible for 20.6% of fatal accidents in the workplace encountered in the European Union (EU) [6], whereas it engaged only 10% of the working population [7]. The overview of fatalities in the United States of America (USA) revealed that 4339 employees lost their lives in the construction sector from 2011 to 2015 [8]. Similarly, employees in the United States of America's construction sector experienced 7% of entire private-sector injuries in 2017 [9]. With particular reference to the United States of America, the fatality rate in the construction sector in 2018 was 3.0 per 100,000 employees with 1008 fatal occupational injuries [10]. In Japan, the construction sector's occupational accident rate was 32.2% in 2020 [11]. The figure for fatal injuries in the UK construction sector was 39 in 2020/21, a reduction of 3 from the preceding year with a total figure of 42 [12]. In Australia, the three priority sectors, agriculture, construction, and road transport, accounted for 55% of employee fatalities between 2015 and 2019, with the construction sector accounting for 17% [13].

Furthermore, in terms of monetary value, according to Gibb et al. [14], an estimated sum of £848 M was spent yearly by the UK employers in the construction industry apart from expenditures borne by the individual and the society, while the Health Safety Executive 2019 (HSE 2019) [15] highlighted that a sum of £1.2 billion was spent in the construction industry for injuries and ill-health in 2017/2018. Occupational accidents in the construction industry cost Australia a sum of 2860 ($ million) between 2012 and 2013 [16]. Similarly, Geetha M. Waehrer et al. [17] in 2002 highlighted that the costs of occupational accidents in the United States of America were estimated at $11.5 billion, 15% of these costs were from the private sector, and in general, the cost per case of injury was $27,000, nearly twice the case per cost of $15,000 for the entire sector in 2002. The sum of $4,634,501,000 was the total estimated cost of construction fatalities in the USA from 2011 to 2015, as highlighted by Manzo [8].

Hazardous working conditions influence employees in different methods. High occupational accidents may create fear among employees who intend to be working in such organizations. Therefore, an organization needs to identify the potential risk before the commencement of any construction work. This can be accomplished by assigning competent workers to perform hazard identification and risk assessment. A

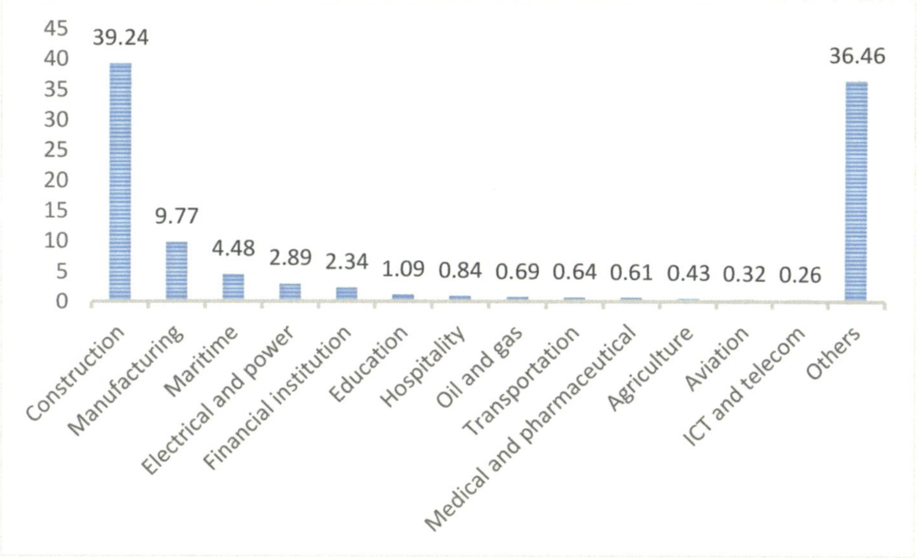

Figure 1.
Accident classification by industry (2014–2016).Source: Federal Ministry of Labour and Employment, Abuja, Nigeria.

vibrant method is required to be defined regarding how hazards are to be identified and monitored to ensure that construction sites are kept free from accidents and injury [18].

Nigerian construction industry is not exempted from the menace of occupational accidents. The construction sector accounted for the highest percentage of occupational accidents that occurred between 2014 and 2016 [19], as illustrated in **Figure 1**. The problems of occupational accidents require drastic steps. Therefore, this study aims to identify the approaches toward improving occupational health and safety in the Nigerian construction industry.

2. Literature review

Improving health and safety in the construction industry remains a priority because the construction sector stands out among all other sectors as the highest contributor to fatal occupational accidents. It was found that the traditional burden of ensuring the health and safety of construction worksites has been placed exclusively on the organizations. Although organizations will always bear the duty for construction site health and safety, the novel perception of prevention through design could be seen to allow Engineers, Architects, and other professionals to contribute to improving site health and safety, likewise the step for the prevention of occupational accidents. Similarly, the responsibility of frontline managers and senior-level leadership is to vigorously work together to reduce constraints on employee participation in the prevention of occupational accidents [20]. According to ILO 2013 [21], managing risks and identifying hazards are important steps to improving health and safety and health at the organizational level. Also, the provision of information and training courses to employees is a legal prerequisite for the prevention of occupational accidents. Managers must identify hazards faced by their employees, likewise employees must recognize the hazards they are exposed to.

According to the study by Bhattacharjee et al. [22], the tactics for improving health and safety in the construction industry include personnel selection, technological intervention, behavior modification, poster campaigns, quality circle, exercise and stress management, safety climate, near-miss accident reporting, and zero injury technique. Mollo et al. [23] highlighted that the process of improving OHS in the construction sector would be accomplished by utilizing learning by doing. According to Chan et al. [24], to guarantee health and safety, and to ensure the sector is more efficient, it is important to take into consideration learning from incidence factors in the construction industry that permits the development of active health and safety measures. Learning by doing permits site managers to transfer skills and knowledge to employees. The study by Osei-Asibey et al. [25] focused on the framework to enable construction participants to recognize problem areas on the construction site and implement the necessary improvement measures to build a healthy and safe construction site. Five expected improvements were identified, namely: (1) qualified participants, (2) positive behavior and attitude, (3) better working conditions and management's commitment, (4) suitable tools and equipment, and (5) better construction health and safety knowledge. According to Manzo [8], four policy approaches to ensure safer and healthy working conditions in the construction sectors are (1) increasing resources for conducting inspections, (2) sustaining or implementing state-acknowledged wage laws, (3) the introduction of responsible bidder ordinances, and (4) avoiding attacks on unions.

Furthermore, according to Kheni et al. [26], robust and well-structured institutional plans to ensure health and safety (H&S) standards in the workplace are required to improve occupational health and safety. The OHS performance of construction sites will remain poor until professionals and owner-managers in the built environment are willing to change their orientations and take up the responsibility. Similarly, Misiurek K. and Misiurek B. [27] highlighted that not technical problems but human errors have the utmost influence on the incidence of occupational accidents. According to Misiurek K. and Misiurek B. [27], the methodology for improving OHS in the construction sector is based on the selected constituents from the training within the industry program which includes: the selection of a procedure to be executed in the methodology, creation of a work breakdown sheet, organizing the procedure for eradicating and improving key points, preparation of warnings for each health and safety-related key point, identification of essential personal protective equipment, preparation of a work breakdown sheet for a new technique for the execution of an operation, and training of workers in the new standard of work and sustaining the methodology. All these factors contribute to the eradication of problems connected with the root causes of human mistakes. Sharma and Kumar [28] highlighted that training within industry programs deals with the following problems: (1) how to convey knowledge efficiently to the employees in the construction sites, how and why should it be done in a definite manner, (2) how to provide training to employees under their requirements, (3) how to improve occupational health and safety by applying a series of questions through the 5W1H method, (4) how to outline a strategy for the building of interactions within the team, and (5) how to use technical processes for the recognizing, eradication, and protection against the hazards.

Colin Fuller and Luise Vassie [29] highlighted that monitoring the application of health and safety management procedures and policies is a vital feature of improving health and safety, as it ensures an organization determines whether recommended standards are followed and are in line with the organizational objectives. Measures to improve occupational health and safety can be accomplished through (1) compliance with health and safety legislation and management system, (2) monitoring deficiencies in health and safety performance, (3) recording performance information that enables preventative and corrective management activities, (4) documentation of the root causes of occupational accidents and nonconformances, and (5) application of quantitative and qualitative measures in preventative and corrective activities.

Antwi-Afari et al. [30] highlighted that with the development of sensing and warning-based innovation, practitioners and researchers have realized that their implementation could offer effective solutions to the improvement of OHS in the construction sector. Improving OHS is a social procedure and the application of economic justification influences the process. Though, organizations and workers are not firmly rational and economically analytical individuals. But, Broek K Van Den et al. [31] pinpoint that occupational accident costing is a valuable method of measuring, managing, and eventually improving health and safety. Mossink and Licher [32] highlighted OHS and profits remain distinct goals, nevertheless the economic theory can give suggestions as to when improvements serve both profits and OHS. Myers et al. [33] maintain those economics can impact decision-makers, but the economic approach is sometimes challenging to defend in front of an audience unwilling to accept health and safety messages, in that circumstance, a narrative approach of incorporating economic arguments works better.

Improvements in occupational health and safety at work can bring economic advantages to organizations, employees, and societies as a whole [34]. Occupational accidents can give rise to serious costs for an organization, for small establishments, in particular, occupational accidents can have substantial financial consequences. British Safety Council 2014 [35] asserts that investing in the management of OHS is an opportunity to gain a variety of advantages including boosting the organization's image and avoiding the costs of penalties. EU-OSHA 2007 [36] highlighted the benefits of investing in OHS, including that it (1) helps to demonstrate that an organization is socially responsible, (2) protects and improves brand image and value, (3) supports maximizing the productivity of workforces, (4) improves workers' commitment to the organization, (5) builds a more competent and healthier workforce, (6) reduces operational costs and disruption of the progress of work, (7) enables organizations to meet customers' expectations on OHS, and (8) encourages the employees to stay longer in active life.

Lingard [37] highlighted that ensuring occupational health and safety knowledge is used to inform decisions made in the planning and design stages of projects remains a problem for the construction industry. The generally cited difficulty is the extent of construction occupational health and safety knowledge owned by the upstream decision-makers. But preconstruction decision-makers must be better informed concerning occupational health and safety to allow them to understand more insightful risk consequences of their decisions. The construction OHS challenge is complex, and it needs a refined answer. Decomposing and solving a single feature of the problem is doubtful to bring about important improvements. Preferably, the incorporation of multiple approaches is essential in the delivery of improvements in the construction occupational health and safety, as well as in the prevention of accidents in the sector.

3. Method

A quantitative method was adopted for this study with a structured questionnaire of 443 distributed to the construction professionals in the Federal Capital Territory, Abuja, Nigeria. To achieve a degree of generalization, the construction professionals' population that was considered were Architects, Quantity Surveyors, Engineers, and Builders in the Federal Capital Territory, Abuja, Nigeria. To ascertain the total number of members of each of the professional bodies, a request was sent to them. The sampling frame is 8974. The number of respondents was calculated at a 95% confidence level with a maximum error of 5%. The sampling size is 443. The questionnaires were proportionally allotted according to the profession through a stratified sampling method. The measurement scale of the questionnaire is illustrated in **Table 1**, in which respondents chose options arranged in a 5-point Likert scale from strongly agree to strongly disagree.

(A) Establishment of the Nigerian Construction Industry Development Board with roles to include	
1	For researching OHS for the prevention of occupational accidents
2	Compilation of occupational accidents' data to identify root causes and for future prevention
3	Monitoring construction professionals' activities concerning OHS
4	Organizing OHS training
5	Development of OHS plan for the country
(B) Technical assistance and collaboration among construction professionals	
1	Support for recognition of OHS for the prevention of occupational accidents through a national action program.
2	Forging, leveraging, promoting knowledge-sharing, and developing policy innovations on OHS.
3	Providing suitable technical assistance to OHS in the development and consolidation of the statutory and institutional framework on health and safety rules for the prevention program.
4	Provision of technical assistance on OHS in construction sites to be supported by industrial-based experience with the tested practice of application of preventive measures and capacity-building on OHS.
5	Technical assistance for the appraisal of health and safety standards before implementation or adoption of new health and safety standards.
(C) Skill development in the management and communication of occupational health and safety	
1	Ensure technology on OHS is adapted to local conditions for the prevention of occupational accidents.
2	The introduction of new technology on OHS should be complemented by adequate information and training.
3	The hazards associated with new technologies used in the construction sites have to be known and effective measures are taken to remove or manage them.
4	Procuring relevant databases and knowledge on OHS from developed countries.

5	Consultation with workers' representatives whenever new technology on OHS is introduced, such technology should not be imported without sufficient measures established, including information on the safe application in the language of the importing nation.
(D) Awareness-raising and advocacy on occupational health and safety	
1	The OHS program should include strategies to promote broad awareness of the social and economic significance of improving working conditions.
2	Better communication and work relationships on OHS in the organization.
3	Occupational health and safety (OHS) awareness campaign targeted at acquainting both employees and management with hazards in the construction sites and their obligations in the prevention of occupational accidents.
4	Occupational health and safety (OHS) education raises awareness and positive attitudes that are conducive to health and safety at work.
5	Communicating skills that will allow managers and employees in the construction sites to identify risk problems give rise to occupational accidents.
(E) Use of International Labour Organization (ILO) mechanism on occupational health and safety	
1	Application of codes of practice on OHS to suit local condition
2	Use of ILO 2001 Occupational Health and Safety Management System (OHSMS)
3	Ratification of ILO conventions
4	Application of ILO policy on OHS
5	Use of ILO standards on OHS to suit local standards.
(F) International collaboration with other professional bodies on health and safety	
1	Team collaboration with international organizations on OHS as the immediate drivers of health and safe work practices
2	Collaborations focus on main areas like setting and benchmarking OHS standards
3	Collaborations with foreign occupational health and safety research organizations and organizing conferences bringing together international experts on OHS
4	Collaboration to better accomplish OHS strategic outcomes
5	Collaboration to explore avenues of raising OHS standards
(G) Proper monitoring and recording of all injuries	

1	Recording performance information that enables preventative and corrective management activities
2	Monitoring deficiencies in health and safety performance
3	Documentation of the root causes of occupational accidents and nonconformances
4	Compliance with health and safety legislation and management system
5	Application of quantitative and qualitative measures.
(H) Adequate allocation of resources (human, financial, and technology) on OHS	
1	Allocation of the resources essential to accomplish and sustain an acceptable risk level on OHS.
2	Assign adequate resources to accomplish the objectives defined in the implementation plans of OHS.
3	The distribution of resources follows logically in harmony with a comprehensive problem-solving procedure on OHS
4	Resources to develop the future direction of the OHS.
5	Occupational health and safety resources for better OHS programs.

Table 1.
Measurement scale of approaches to improving OHS.

3.1 Data analysis

A total number of 399 questionnaires were returned out of 443 distributed. Exploratory factor analysis (EFA) was conducted on the data with SPSS software version 26. The confirmatory factor analysis (CFA) was used to validate the factors obtained from EFA. The CFA was conducted through structural equation modeling-AMOS. Some indices were used to evaluate the goodness-of-fit index of the CFA model. These indices include root means square error of approximation (RMSEA), the goodness-of-fit index, adjusted goodness-of-fit index, comparative fit index (CFI), and normal fit index [38]. SPSS software version 26 was used for descriptive analyses.

4. Results

4.1 Respondents' demography information

Four hundred forty-three structured questionnaires were distributed, and 399 were returned representing 90.07%. The respondents' background information is as follows. The academic qualifications: National Diploma (3%), Bachelor of Science/Higher National Diploma (64%), Postgraduate Diploma (10%), Master of Science (17%), and Ph.D. (6%). Professional affiliation: Architects (20.8%), Builders (3.3%), Engineers (58.6%), Quantity Surveyors (16.8%), and Surveying and Geoformation (0.5%). Year of experience: 1–5 years (7.85%), 6–10 years (8.5%), 11–15 years (22.8%), 16–20 years (27.8%), and 21 years and above (33.1%). Firms, organizations, and institutions: Consulting firms (25.3%), Contracting firms (16.8%), Developer organizations (23.1%), Educational institutions (6.8%), Governmental agencies (22.8%), and Multinationals (5.3%).

4.2 Dimensionality of the scale

Evaluation of the data was conducted through EFA. The cutoff value of the exploratory factor was fixed at 0.40 according to Yong and Pearce [39], through the principal component in varimax rotation. Values less than 0.4 were dropped, but values above 0.4 were considered for the analysis. **Table 2** shows the result of the EFA for this study. Kaiser-Meyer-Olkin (KMO) was used to evaluate the appropriateness of data for EFA. Pallant [40] highlighted that a KMO value of 0.6 is suggested as the minimum value for good factor analysis. The KMO result obtained was 0.828, therefore, the data were considered acceptable for the analysis. Confirmatory factor analysis was conducted with structural equation modeling-AMOS. CFA was used to develop and check the psychometric validity of the approaches to the improvement of occupational health and safety in the Nigerian construction industry. The unidimensional approach of OHS is the Establishment of the Nigerian Construction Industry Development Board, technical assistance and collaboration, skill development, awareness-raising and advocacy, use of International Labour Organization mechanism, international collaboration, proper monitoring and recording, and adequate allocation of resources. The initial goodness of fit showed that the model was not fitted as illustrated in **Figure 2**. The result of the indices was CFI = 0.874, incremental fit index (IFI) = 0.875, Tucker-Lewis index (TLI) = 0.864, P = 0.000, RMSEA = 0.05, P = 0.00, and ratio = 2.001. Although RMSEA = 0.05, P = 0.00, and ratio = 2.001 met the minimum threshold, CFI = 0.874, IFI = 0.875, and TLI = 0.864 did not meet the minimum threshold. Thereafter, variables that contributed to the poor fit of the model were dropped [41]. The total number

Rotated component matrix[a]								
	Component							
	1	**2**	**3**	**4**	**5**	**6**	**7**	**8**
C1Q1		.687						
C1Q2		.790						
C1Q3		.802						
C1Q4		.730						
C1Q5		.682						
C2Q1	.635							
C2Q2	.675							
C2Q3	.795							
C2Q4	.812							
C2Q5	.776							
C3Q1			.731					
C3Q2			.690					
C3Q3			.790					
C3Q4			.622					
C3Q5			.670					
C4Q1					.612			
C4Q2					.758			

Rotated component matrix[a]								
	Component							
	1	2	3	4	5	6	7	8
C4Q3					.649			
C4Q4					.723			
C4Q5					.608			
C5Q3								.704
C5Q4								.700
C5Q5								.729
C6Q1				.545				
C6Q2				.686				
C6Q3				.628				
C6Q4				.619				
C6Q5				.710				
C7Q1						.735		
C7Q2						.655		
C7Q3						.551		
C7Q4						.608		
C7Q5						.570		
C8Q1							.583	
C8Q2							.536	
C8Q3							.734	
C8Q5							.640	

Extraction method: Principal component analysis.
Rotation method: Varimax with Kaiser normalization.[a]

[a]*Rotation converged in seven iterations.*

Table 2.
EFA of approaches to improving OHS.

of six variables with low factor loadings was eliminated. **Figure 3** illustrates the final model with acceptable goodness of fit. **Table 3** illustrates the acceptable model fit of the approach to improving occupational health and safety of the Nigerian construction industry with the following indices such as minimum discrepancy divided by degree of freedom (CMIN/DF) = 1.846, CFI = 0.912, standardized root mean squared residual (SRMR) = 0.063, RMSEA = 0.055, and Pclose = 0.167. **Table 4** illustrates the standardized regression weight of the factors which are all above the minimum threshold of 0.5 of a good model.

4.3 Reliability and validity of approach to improving OHS

Table 5 illustrates that the Cronbach coefficient is greater than 0.6, therefore it is necessary to measure the adequacy level of the reliability to test the causal effect of the relation. According to Ursachi et al. and Nunnally and Bernstein [42, 43], the

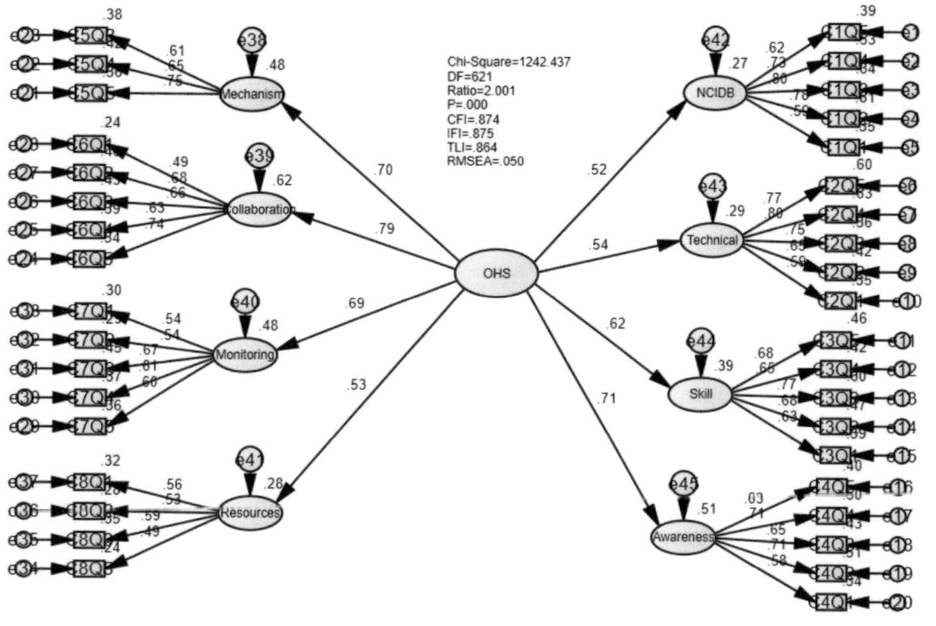

Figure 2.
Initial model of the approaches to improving OHS.

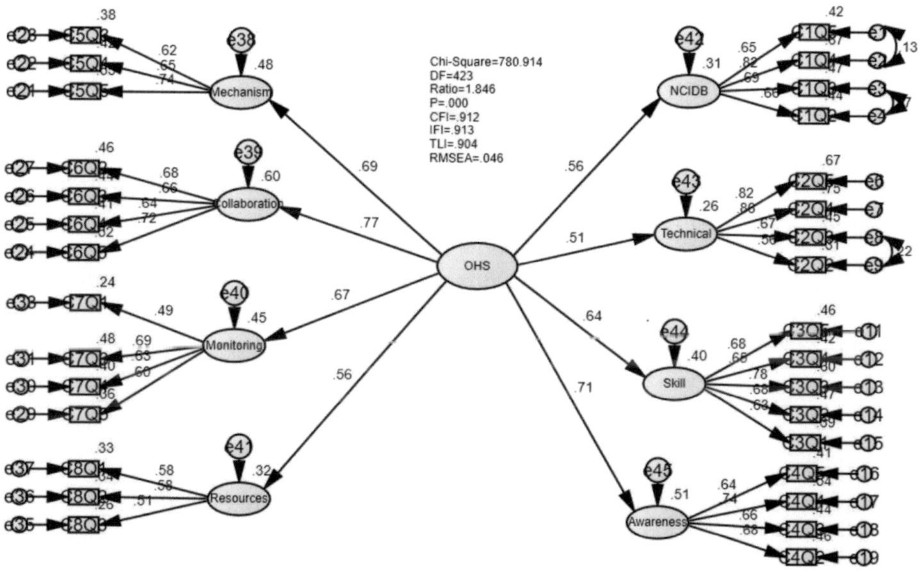

Figure 3.
Final model of the approaches to improving OHS.

generally accepted standard is Cronbach's alpha of 0.6–0.7 which shows a suitable level of reliability. Also, 0.8 or higher is a good level. But, values greater than 0.95 are not certainly good, as they can be seen as a sign of redundancy [44]. Similarly, the composite reliability in all cases exceeded the least level of 0.6 as suggested by Bagozzi and Yi [45]. While the average variance explained was equal to 0.5 according to Awang [46], however, Bagozzi [47] stated that a value of 0.3 and above could be

Name of the index	Level of acceptance	Results obtained
Pclose	<0.05	0.897
RMSEA	<0.08	0.046
CFI	> 0.90	0.912
IFI	> 0.90	0.913
SRMR	< 0.08	0.055
Ratio	2–5	1.851

Table 3.
Goodness of fit of the model of approaches to improving OHS.

Construct	Path	Construct	Factors
NCIBD	<---	OHS	.557
Resources	<---	OHS	.565
Monitoring	<---	OHS	.673
ILO mechanism	<---	OHS	.694
Awareness-raising and advocacy	<---	OHS	.713
Skill development	<---	OHS	.635
Technical assistance	<---	OHS	.510
International collaboration	<---	OHS	.774

Table 4.
Standardized regression weight of the factors of approaches to improving OHS.

considered. **Figure 2** indicates that the model fit is acceptable. Also, the standardized regression weights of the success factors were higher than 0.5 and significant at the 95% confidence level, confirming the convergent validity of the constructs as illustrated in **Table 5**. The validity and reliability of the approaches to improving occupational health and safety of the Nigerian construction industry had been demonstrated and structured in 31 variables at eight dimensions.

5. Discussion

The study identified the approaches through which the occupational health and safety of the Nigerian construction industry can be improved. These include the Establishment of the Nigerian Construction Industry Development Board (NCIDB), technical assistance and collaboration, skill development, awareness-raising and advocacy, use of International Labour Organization mechanisms, international collaboration, proper monitoring and recording, and adequate allocation of resources. The validity of the approach was tested and found to be acceptable. Confirmatory factor analysis was used to validate exploratory factor analysis-derived factor structures of the measurement scale [48]. The yardsticks of the different model fit indicators were considered. RMSE values of 0.046 in this study illustrate a good fit, as suggested by Bentler and Bonett [49] that RMSEA values of 0.05 and 0.08 are satisfactory. The CFI and IFI values in the model were

Code	Factor loading	Composite reliability	Average variance explained	Cronbach Alpha
Establishment of the Nigerian Construction Industry Development Board				
C1Q5	0.652	0.8	0.5	0.83
C1Q4	0.821			
C1Q3	0.688			
C1Q2	0.662			
Technical assistance and collaboration among construction professionals				
C2Q5	0.818	0.8	0.5	0.84
C2Q4	0.863			
C2Q3	0.674			
C2Q2	0.561			
Skill development in the management and communication of occupational health and safety				
C3Q5	0.678	0.8	0.5	0.81
C3Q4	0.65			
C3Q3	0.775			
C3Q2	0.682			
C3Q1	0.625			
Awareness-raising and advocacy on occupational health and safety				
C4Q5	0.642	0.8	0.5	0.82
C4Q4	0.735			
C4Q3	0.663			
C4Q2	0.676			
Use of International Labour Organization mechanism on occupational health and safety				
C5Q5	0.742	0.7	0.5	0.71
C5Q4	0.65			
C5Q3	0.616			
International collaboration with other professional bodies on health and safety				
C6Q5	0.724	0.5	0.8	0.77
C6Q4	0.639			
C6Q3	0.663			
C6Q2	0.677			
Proper monitoring and recording of all injuries				
C7Q5	0.598	0.4	0.7	0.70
C7Q4	0.629			
C7Q3	0.69			
C7Q1	0.494			
Adequate allocation of resources (human, financial, and technology) on OHS				
C8Q3	0.514	0.3	0.6	0.60
C8Q2	0.583			
C8Q1	0.576			

Table 5.
First-order CFA of the approaches to improving OHS.

higher than 0.9, therefore illustrating a reasonably good fit [49]. Generally, indices values support the concept of a good fit and offer confirmatory proof for the factor structure. In the appraisal of the psychometric properties of the approaches to improving occupational health and safety of the Nigerian construction industry, construct validity was supported by CFA and EFA. From the standardized loading factors of the construct, international collaboration has the highest value of.774. According to Fingerhut [50], the exchange of experience and collaboration among centers is assisted by meetings of collaborating centers, which can support institutional capacity in occupational health and safety in regions and nations. There are great benefits in international collaboration where knowledge and experiences are shared and where we learn from the challenges and strengths faced by different nations in the quest for healthier and safer construction sites. Similarly, Hsu et al. [51] point out that collaboration could act as the instant driver of health and safe work practices. This sharing of experiences and information can help to improve occupational health and safety in the spirit of solving OHS matters in an informed manner.

Furthermore, from the standardized loading factor, awareness-raising and advocacy have the second-highest value of.713. There is a need for consideration of awareness of occupational health and safety and advocacy for a greater awareness of hazards in the workplace, which should include both health and safety programs. Employees and management team participation in health and safety programs allows them to easily accept and understand OHS changes. According to Pillay et al. [52], the absence of awareness in OHS contributes to construction accidents. Trethewy [53] highlighted that practical results and positive feedback produced from OHS observation and reporting not only improve awareness of OHS behavior and workplace safety but also induce a change process that assists to renew organizational safety culture and values. According to Teo et al. [54], technical and safety training can reduce the absence of health and safety awareness inherited in construction employees.

Use of ILO mechanism, proper monitoring, and recording of occupational accidents and skill development each has the following loadings of.694, .673, and.635, respectively. ILO has produced many instruments on OHS that can help to improve OHS. The need to use these instruments is important for improving the OHS of the Nigerian construction industry. While Yiu et al. [55] advocated for a comprehensive occupational health and safety monitoring system to be established to improve the success of safety management. Also, many nations are making important investments in training, education, job creation, and skill development, to this extent, ILO 2018 [56] suggests that it is significant to include OHS in these programs. To do so, it is essential to understand and solve the problems of health and safety risks confronted by employees and management in the workplace.

Other approaches with their loading factors are the Establishment of the Nigerian Construction Industry Development Board-.557, adequate allocation of resources-.565, and technical assistance from the professionals. The need for the establishment of a board in the construction industry similar to the Construction Industry Development Board (CIDB) in Malaysia and Reporting of Injuries, Diseases and Dangerous Occurrences Regulations (RIDDOR) in the UK will support the Nigerian construction industry in improving the occupational health and safety of the sector. While Manuele [57] pointed out that it is absolute that if sufficient resources are not provided on OHS, tolerable risk levels cannot be sustained.

6. Conclusions

The construction industry is universally ranked as the most hazardous sector. However, many nations have put immense efforts to improve health and safety performance as well as bring about the reduction in injuries' occurrence. Nevertheless, the construction industry remains to lag behind other sectors with a high record of accidents. The construction industry must be stimulated to standardize its safety and health practices and performances. Better opportunities must be engaged to learn from failures with the application of preventive measures to curb accident occurrence. World has become a global village, through crossing borders-arrangements, technology, and corporation. Construction worker health and safety have become a concern that is shared universally. Since construction health and safety issues are very alike from nation to nation, they can therefore be addressed on a global scale. Solutions to health and safety issues in one nation can readily be implemented in other nations to produce further improvements. Improving occupational health and safety will contribute to the accomplishment of the Sustainable Development Goal (SDG) on decent work and improved economic growth of the Nigerian construction industry's global cooperation. The study recommends that government at all levels should participate in the stringent legal enforcement of occupational health and safety legislation. Also, the economic approach to OHS must be viewed from a wider perspective, not only focusing on just economic benefits and costs. It is essential to outline the potential worth of OHS as seen from a business viewpoint. Achieving greater coherence in OHS could come through well-defined regional and country initiatives.

Author details

Kamoli Adetunji*, Mohd Saidin Misnan, Mohamad Zahierruden Ismail,
Farah Nadiah Abdul Rahim and Zulkiflee Abdul-Samad
Faculty of Built Environment and Surveying, Department of Quantity Surveying,
Universiti Teknologi Malaysia, Johor, Malaysia

*Address all correspondence to: qsakinadetunjik@gmail.com

References

[1] Alhelo AA, Alzubaidi R, Rashid H. A framework supporting health and safety practices in the United Arab Emirates ' construction projects. Sustainability. 2023;**15**:1-14

[2] Dine G, Reed S, Oosthuizen J, Masaka E. Identifying occupational health and safety risks among environmental health officers in Australia and New Zealand through an online survey. Medicine (Baltimore). 2023;**102**:e33270

[3] Zhou Z, Li Q, Wu W. Developing a versatile Subway construction incident database for safety management. Journal of Construction Engineering and Management. 2012;**138**:1169-1180

[4] Rowlinson S. Construction safety management: The case for a new approach to research-informed change. In: A Research Agenda for Construction Management. Cheltenham, UK: Edward Elgar Publishing Ltd; 2023. pp. 1-314

[5] ILO. Construction: A hazardous work. ILO. 2015;**2015**:1-2

[6] Eurostat 2021 Accidents at work statistics: Statistics explained Eurostate European Union 20211-7

[7] Rowlinson S. Construction Safety Management Systems. London, UK: Spon Press; 2004

[8] Manzo J. The $ 5 Billion Cost of Construction Fatalities in the United States: A 50 State Comparison. United States of America: Midwest Economic Policy Institute; 2017

[9] AFL-CIO. A National and State-By-State Profile of Worker Safety and Health in the United States. Washington, DC; 2019

[10] Bureau of Labor Statistics. National Census of Fatal Occupational Injuries in 2018. United States of America: U.S Department Labor; 2019

[11] Japan Industrial Safety and Health Association. Industrial Accidents in CY2020 in Japan (Fixed Data) April 2021 Translated by JISHA (Data Sources) Safety Division. Japan: Industrial Safety and Health Department, Labour Standards Bureau, Ministry of Health, Labour and Welfare; 2021

[12] Health and Safety Executive. Workplace fatal injuries in Great Britain 2020. Health Safety. Executive. 2021;**2121**:1-19

[13] Safe Work Australia. Work-related Traumatic Injury Fatalities. Australia; 2019

[14] Gibb A, Drake C, Jones W. Costs of occupational ill-health in construction. England: Loughborough University; 2018

[15] HSE. Costs to Britain of workplace fatalities and self-reported injuries and ill health, 2017/1. Health Safety Executive. 2019;**18**:1-30

[16] Safe Work Australia. The cost of work-related injury and illness for Australian employers, workers and the community: 2012-2013. System. 2015;**2015**:1-48

[17] Waehrer GM, Dong XS, Miller T, Haile E, Men Y. Costs of occupational injuries in construction in the United States. Natlional Institution on Health. 2007;**39**:1258-1266

[18] Nkqayana Y, Smallwood J. Health and safety coordination between main contractors and subcontractors on a power station project. In: Digital Transformation of Health and Safety in Construction. Porto, Portugal; 2023. pp. 339-347

[19] Federal Ministry of Labour and Employment. Nigeria Country Profile on Occupational Safety and Health. Abuja, Nigeria: Federal Ministry of Labour and Employment; 2016

[20] Haas EJ, Cauda E. Using Core elements of health and safety management systems to support worker well-being during technology integration. International Journal of Environmental Research and Public Health. 2022;**2022**:19

[21] ILO. Safety and Health at Work: Hopes and challenges in development cooperation *Ex. an EU-ILO Jt. Proj. "Improving Saf. Heal. Work through a Decent Work Agenda* 1-25. 2013

[22] Bhattacharjee S, Ghosh S. Safety improvement approaches in construction industry: A review and future directions. Omaha, NE, United States of America: University of Nebraska-Lincoln; 2011

[23] Mollo LG, Emuze F, Smallwood J. Improving occupational health and safety (OHS) in construction using training-within-industry method. Journal of Financial Management of Property and Construction. 2019;**24**:655-671

[24] Chan APC, Guan J, Choi TNY, Yang Y, Wu G. Improving safety performance of construction workers through learning from incidents. International Journal of Environmental Research and Public Health. 2023;**20**:1-26

[25] Osei-Asibey D, Ayarkwa J, Acheampong A, Adinyira E, Amoah P. Framework for improving construction health and safety on Ghanaian construction sites. Journal of Building Construction Planning Research. 2021;**9**:115-137

[26] Kheni NA, Gibb AGF, Dainty ARJ. The management of construction site health and safety by small and medium-sized construction businesses in developing countries: A Ghana case study. In: Proceedings 22nd Annual ARCOM Conference, 4-6 September 2006. Vol. 1. Birmingham, UK. Association of Researchers in Construction Management; 2006. pp. 273-282

[27] Misiurek K, Misiurek B. Methodology of improving occupational safety in the construction industry on the basis of the TWI program. Safety Science. 2017;**92**:225-231

[28] Sharma A, Kumar P. Strategies for improving occupational safety in construction industries. Jpurnal of Information Computer Science. 2020;**10**:1568-1574

[29] Colin WF, Luise HV. Health and Safety Management: Principles and Best Practice. 1st ed. England: Pearson Education Limited; 2004

[30] Antwi-Afari MF, Li H, Wong JKW, Oladinrin OT, Ge JX, Seo JO, et al. Sensing and warning-based technology applications to improve occupational health and safety in the construction industry: A literature review. Engineering Construction Architect Management. 2019;**26**:1534-1552

[31] Broek K Van Den, Greef M De, Heyden S Van Der, Kuhl K, Schmitz-Felten E, Van den Broek K, De Greef M and Van der Heyden S 2011 Socio-economic costs of accidents at work and work- related ill health EU Public. 2011217

[32] Mossink J, Licher F. Costs and benefits of occupational health and safety. In: European Conference on the

Costs and Benefits of Occupational Safety and Health. Dublin, Ireland: European Foundation for the Improvement of Living and Working Conditions; 1997. pp. 1-11

[33] Myers M, Cole H, Mazur J, Isaacs S. Economics & Safety Understanding the cost of injuries and their prevention. Safety. 2008;**53**:14-15

[34] EU-OSHA. Inventory of Socioeconomic Costs of Work Accidents. Luxembourg: European Agency for Safety and Health at Work; 2002

[35] British Safety Council. The business benefits of health and safety: A literature review. London, United Kingdom: British Safety Council; 2014

[36] EU-OSHA. The business benefits of good occupational safety and health. European Agency Safety Health Work. 2007;**2007**:1-2

[37] Lingard H. Occupational health and safety in the construction industry. Construction Management Economy. 2013;**31**:505-514

[38] Byrne BM. Structural Equation Modeling with AMOS: Basic Concepts, Applications, and Programming. NY, New York: Routledge Taylor & Francis Group; 2016

[39] Yong AG, Pearce S. A Beginner's guide to factor analysis: Focusing on exploratory factor analysis. Quantum Methods Psychology. 2013;**9**:79-94

[40] Pallant J. SPSS Survival Manual (Two Penn Plaza). New York, NY: Open University Press; 2016

[41] Baumgartner H, Steenkamp JBEM. Exploratory consumer buying behavior: Conceptualization and measurement. International Journal of Research Marking. 1996;**13**:121-137

[42] Ursachi G, Horodnic IA, Zait A. How reliable are measurement scales? External factors with indirect influence on reliability estimators. Procedia Economic Finance. 2015;**20**:679-686

[43] Nunnally JC, Bernstein IH. Psychometric Theory. United States of America: McGraw-Hill, Inc.; 1994

[44] Streiner DL. Starting at the beginning: An introduction to coefficient alpha and internal consistency. Journal of Personality Assessment. 2003;**80**:99-103

[45] Bagozzi RP, Yi Y. On the evaluation of structural equation models. Journal of the Academy of Marketing Science. 1988;**16**:74-94

[46] Awang ZH. A Handbook on Structural Equation Modelling. MPWS Rich Publication; 2015

[47] Bagozzi RP. Evaluating structural equation models with unobservable variables and measurement error: A comment. Journal of Marketing Research. 1981;**18**:375-381

[48] Kyriazos TA. Applied psychometrics: Sample size and sample power considerations in factor analysis (EFA, CFA) and SEM in general. Psychology. 2018;**09**:2207-2230

[49] Bentler PM, Bonett DG. Significance tests and goodness-of-fit in analysis of covariance structures significance tests and goodness of fit in the analysis of covariance structures. Psychological Bulletin. 1980;**88**:5SS – 606

[50] Fingerhut M. The global occupational health network. Global Occupational Health Network. 2004;**2004**:1-12

[51] Hsu SH, Lee CC, Wu MC, Takano K. The influence of organizational factors on safety in Taiwanese high-risk industries. Journal of Loss Prevention in the Process Industries. 2010;**23**:646-653

[52] Pillay J, Ryan R, Charles M. Guide to Best Practice for Safer Construction: Literature Review 'from Concept to Completion'. Brisbane, Australia: Icon. Net Pty Ltd; 2007

[53] Trethewy RW. OHS performance: Improved indicators for contractors. Journal of Occupational Safety and Health. 2000;**16**:527-534

[54] Teo EAL, Ling FYY, Ong DSY. Fostering safe work behaviour in workers at construction sites. Engineering Construction Architect Management. 2005;**12**:410-422

[55] Yiu NSN, Sze NN, Chan DWM. Implementation of safety management systems in Hong Kong construction industry – A safety practitioner's perspective. Journal of Safety Research. 2018;**64**:1-9

[56] ILO. Improving the Safety and Health of Young Workers. International Labour Organization; 2018

[57] Manuele FA. An Overview of the Occupational Health & Safety Management Systems Standard Prof. Safety

Chapter 7

Risk of Delay in Construction Projects

A.M. Faten Albtoush

Abstract

The construction sector is considered the most important sector, as it contributes to raising the country's economy. Therefore, it is necessary to search for the obstacles. One of these obstacles is the delay, as the project is rarely completed within the planned period. Among the most prominent reasons for this delay are change orders, financial problems for the owner, and slow decision-making on the part of stakeholders. This delay led to various risks, like financial risks, which were represented in exceeding the cost of the project as well as the other in operating the facility and benefiting from the planned benefit. The risk identification process starts with Existence of uncertainty, Preliminary checklist, Risk events consequence scenarios, Logical categorization scheme, Risk mapping, and finally is risk category summary. Therefore, these risks should be analyzed during the planning stage, which can reduce risks during the implementation of the project must be developed. Also, money must be involved in the planning process to be able to fully understand the owner's perception of the project. The provision of the required budget for the project in the early stages ensures the availability of financial liquidity and thus does not delay payment to the contractor.

Keywords: construction projects, delay, risk, stockholders, cost overrun

1. Introduction

Nowadays the construction industry is considered an important sector among the other sectors in any country around the world. Therefore, great attention was given to this sector by several researchers to improve its effectiveness and addressed its problems. One of the challenges facing construction projects worldwide is the delay in completing the required work within the specified time. It either goes past the deadline set forth in the contract or the one agreed upon by the parties for project delivery [1]. On the other hand, time, according to Majid [2], is the amount of time required to complete a building job. When a project is delayed, it indicates that the work cannot be finished by the original contract deadline [2]. The problem of delays in construction projects has become inherent in almost every project, as it is rarely completed within the time specified in the contract. So, because of the nature of construction projects, which distinguishes them from other sectors, they are considered huge in production and need long periods of implementation. However, there are a lot of anticipated and unforeseen factors that can impact the production rate [1].

The delay in the completion of the project does not only affect the stakeholders of the projects, but it extends to be reflected in the impact on other economic fields. Therefore, it is necessary to identify the risks of this delay as a starting point for searching for appropriate solutions to mitigate these risks.

2. Delay in construction projects

Delays in construction projects are caused by several variables. According to Desai and Bhatt, a construction delay is a time overrun in which the project's completion date is later than the date stipulated in the contract or takes longer to finish than the date on which the parties originally agreed to deliver the project [3]. Delays in construction projects are now becoming a frequent problem in most projects due to various reasons that vary from one project to another. Regardless of the reasons for the delay or who is at fault, project delays decrease the contract's value [1]. Therefore, the project manager is crucial in ensuring that the projects are finished within the budgeted time and cost, as stipulated in the contract because delays in construction projects create stress and worry for all the parties under the terms of the contract [4]. There are various types of delays in construction projects, which are based on the source or the affected parties for this delay. For example, construction delays can be categorized into critical delays, non-critical delays, causable delays, non-causable delays, and concurrent delays according to Thorat et al. [5]. According to Ahmed et al. [6], delays may be non-excusable, excusable (with or without compensation), or concurrent. Non-excusable delays are those that are brought on by the contractor, a subcontractor, or another party but are within the contractor's control [6]. Based on the various studies, the major types of delay in construction projects can be summarized into three types, which are excusable delay, non-excusable delay, and concurrent delays, as shown in **Figure 1**.

2.1 Excusable delay

It is the kind of delay that happens as a result of an unforeseen circumstance that neither the contractor nor the subcontractor can control. Excusable delays are caused by the following occurrences: floods, fires, acts of God, general labor strikes, typically harsh weather, etc. [1].

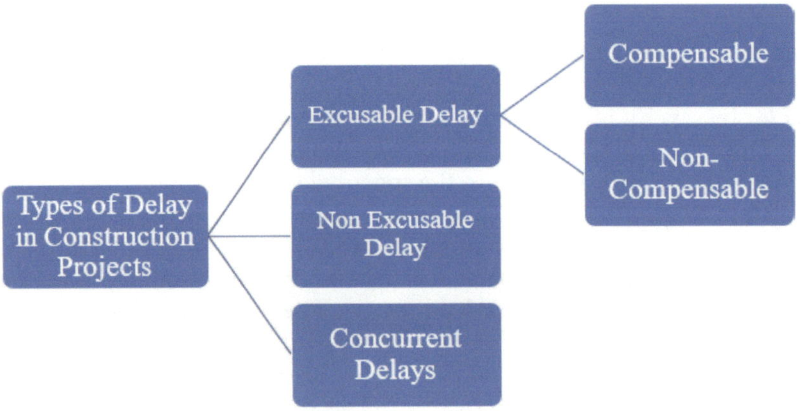

Figure 1.
Types of delay in construction projects.

There are two types of excused delays: compensable and non-compensable. Owner or owner's agents are to blame for compensable delays. Non-compensable delays, however, are brought on by events or persons outside the control of the contractor, the owner, or both. These delays are frequently referred to as "acts of God" because no one is to blame for them [7, 8].

2.2 Non-excusable delay

The foreseeable incident that is within the contractor's control constitutes a non-excusable delay. Examples of this type of delay include late material deliveries from suppliers, late work completion by subcontractors, incorrect work execution by the contractor or subcontractor, a shortage of labor and equipment, and strikes by workers because the contractor intends to stop paying them on time [1].

2.3 Concurrent delays

Concurrent delays are more difficult and common in building projects. This circumstance occurs when multiple factors cause the project to be delayed simultaneously or over a period of time [9].

3. Causes of delay

Many studies have found multiple reasons for project delays, from which many projects in various countries, whether developing or developed, have suffered. Where the causes and the extent of their impact varied according to economic, political, and social factors. For example, even though Malaysia is a nation that is noted for its rapid development, there are still times when projects are delayed. About 17.3% of Malaysia's 417 government projects in 2005 were deemed sick, meaning they had been postponed for more than 3 months or abandoned. One of the industries that significantly contributes to Malaysia's economic expansion is the building industry. However, some of the reasons for and outcomes of construction project delays might be classified as nation-specific [10, 11].

On the other hand, Long et al. [3] discovered that several issues arise in Vietnam throughout the execution of a building project, with schedule and cost overruns being the two main concerns. Another study in Saudi Arabia, Assaf et al. [12] conducted research about construction project delays in different types of projects in the state. It was concluded that 70% of projects experience time overruns [4]. **Table 1** summarizes the most causes of delay in construction projects that are related to different countries.

It is clear in **Table 1** that there are several reasons for the delay in construction projects located in different countries. With the reasons differing from one project to another due to the difference in size, location, type, scope, and requirements. It also differs from one country to another due to the difference in economic, social, and political conditions.

The project manager is crucial in ensuring that the projects are finished within the budgeted time and cost, as stipulated in the contract because delays in construction projects create stress and worry for all the parties under the terms of the contract [4].

According to a study by Rahsid et al. [15], classification in his research is one of the project participants, such as the contractor, the owner, or the consultant, may be

Country	Causes of delay	Study
Saudi Arabia	The causes related to labor, contractor, project, owner, and consultant	[12]
Egypt	financing by contractor during construction, delays in contractor's payment by owner, design changes by owner or his agent during construction, partial payments during construction, and non-utilization of professional construction/contractual management	[13]
Libya	Improper planning, lack of effective communication, design errors, shortage of supply, i.e., steel, concrete, slow decision making, financial issues, shortage of material, cash-flow problems during construction	[7]
Malaysia	Contractor's improper planning, contractor's poor site management, inadequate contractor experience, inadequate client's finance and payments for completed work, problems with subcontractors, shortage in material, labor supply, equipment availability, and failure, lack of communication between parties, and mistakes during the construction stage	[10]
Jordan	Poor design and carelessness of the owner, change orders, weather conditions, site conditions, late delivery, economic conditions, and increase in quantities	[14]
Punjab-Pakistan	Factors related to contractor, client, consultant, material, and equipment	[15]

Table 1.
The main causes of delay in construction projects in different countries.

at fault for the delay. On the other hand, the causes may be related to factors within the project such as delays in the supply of materials, poor management of the site, and difficulty in moving within the project [14]. The delay may also occur as a result of poor relations between the parties involved in the project, whether it is between the owner and the contractor or the contractor and the consultant. Also, it may be a bad relationship between the contractor and his staff on site.

These various reasons clearly affect the success of the project in completing within the required standards related to time, cost, and quality. The impact of the delay in completing the project was not limited to the project only but extended to all other ramifications around the project.

4. Risk of delay

The construction project sector is considered one of the vital sectors that contribute to raising the efficiency of the economy in general. Despite this, it is considered one of the sectors most exposed to various risks. One of these risks is the delay in the completion of projects, the effects of which appeared on the project in particular, as well as its stakeholders, which later extended to the entire sector, in addition to other sectors related to this sector. There are various types of risks in construction projects namely: financial risk, time-related risk, technical risk, market risk, nature risk, external risk, and risk related to the human factor and workplace safety [16]. Sources of construction hazards are listed by Cohen and Palmer as changes in project scope and requirements, design flaws and omissions, poorly defined roles and duties, a lack of qualified employees, force majeure, and new technology [17].

Therefore, during the planning phase of the project life cycle, it is important to provide risk analysis. The results of the risk analysis at this stage will help to reduce the different types of risks that occur during the construction phase. This analysis should be the responsibility of the main stakeholders in the project. However, choosing the appropriate method to respond to risks is an important management process of risk analysis.

The research done by Szymański [16] gave examples of selected situations of reactions on how the investor, contractor, or interested party in a specific situation progresses when making judgments on developing risks, namely: avoiding risk, risk mitigation, dispersion of risk, and absorbing risk.

The factors that cause construction delays that are frequently challenging to fix are time, cost, scope of work, material delivery, benefits, process, resources, stakeholders, or organization [18]. Therefore, when performing their various duties, project team members, in particular customers, consultants, and contractors, should eliminate or mitigate delays [19]. According to a study by Al-Bahar and Crandall [20] the risk identification process is summarized in **Figure 2**.

Tight project schedules, design variations, lengthy approval processes in administrative government departments, client variations, incomplete approval and other documents, inappropriate construction program planning, and insufficient program scheduling are time-related risks identified by Zou et al. [21] that have an impact on project delivery [21].

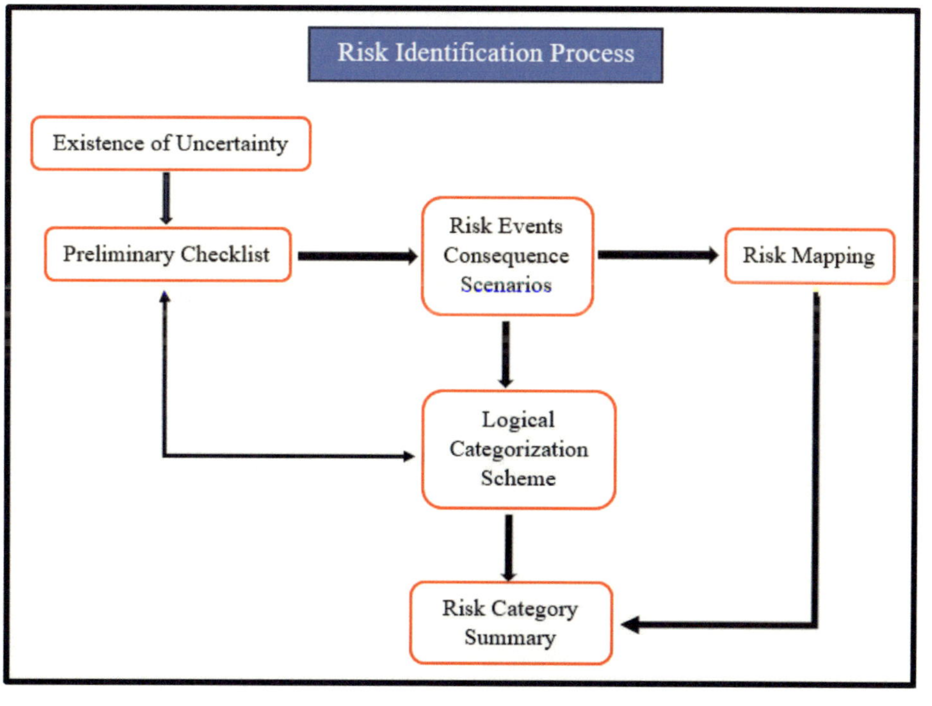

Figure 2.
Risk identification process framework.

5. Effect of delay on construction projects

The effect of delay in construction projects is clearly visible to the parties concerned. For the contractor, the delay incurs additional administrative expenses, which were not taken into account in the initial stages of the project. On the other hand, the delay also delays the owner's use of the project and its operation, and thus deprives the owner of the expected profits at an earlier time. Zhu asserts that the discrepancy between a project's allocated projected cost and actual cost is what causes cost overruns or escalations [22]. Other impacts, like project time overruns, project cost overruns, project abandonment, and litigation, were also addressed [22].

According to Sambasivan and Soon [10]'s research, there are six ways that delays affect how well projects perform in Nigeria. These consequences include conflicts, arbitration, litigation, time and expense overruns, and complete abandonment. The main repercussions of delays and interruptions, however, include time overruns, cost overruns, poor social impact, idle resources, and disagreements [17]. According to Chileshe and Berko, delays in monthly payments to contractors, variances, inflation, and schedule slippage are the main drivers of cost overruns in the Ghanaian road-building industry. Once more, these describe the reasons for delays and the consequences of cost overruns [23].

6. Conclusions

The delay in completing the project within the agreed period is one of the most prominent problems that many construction projects in various countries around the world suffer from. Therefore, the current study focused on identifying the risks of delay. Where the results of the research showed that the effect of delay is not only limited to the project itself but also affects all parties associated with the project, whether directly or indirectly. One of the most prominent risks of delay is the cost overrun resulting from the requirements of this delay, as any additional time requires additional expenses associated with additional labor and materials. It was found that the reasons for the delay were related to the owner, as he was responsible for the early stages of the project. Therefore, more attention must be paid to the planning stage and the preparation of project documents, since most of the reasons were the result of a lack of planning and preparation of project documents. Accordingly, sufficient time must be given for this stage, in addition to selecting qualified specialists to prepare the project documents necessary for its implementation within the requirements and standards specified by the owner. On the other hand, the study recommended the establishment of other studies that focus on identifying potential risks in the planning phase of the project life cycle.

DOI: http://dx.doi.org/10.5772/intechopen.112594

Author details

A.M. Faten Albtoush
Jadara University, Irbid, Jordan

*Address all correspondence to: f.albtoush@jadara.edu.jo

References

[1] Tawfek AM, Bera DK. Delay in construction projects: Types, causes and effects. Project Management Practices, Chief Editor. 2018. pp. 184-192

[2] Majid IA. Causes and Effect of Delays in Aceh Construction Industry. Malaysia: Masters of Science in Construction Management: Universiti technologi Malaysia; 2006

[3] Desai M, Bhatt R. Critical causes of delay in residential construction projects: Case study of Central Gujarat region of India. International Journal of Engineering Trends and Technology. 2013;**4**(4):762-768

[4] Doraisamy SV, Akasah ZA, Yunus R. An overview on the issue of delay in the construction industry. In: CIEC 2014: Proceedings of the International Civil and Infrastructure Engineering Conference 2014. Singapore: Springer; 2015. pp. 313-319

[5] Thorat S, Khandare MA, Kanase AK. Identifying the causes and effects of delay in residential projects. International Research Journal of Engineering and Technology (IRJET). 2017;**4**:2993-2996

[6] Ahmed SM, Azhar S, Castillo M, Kappagantula P. Construction delays in Florida: An empirical study. Final report. Department of Community Affairs, Florida, US. 2002

[7] Saleh AHT, Abdelnaser O, Abdul HKP. Causes of delay in construction industry in Libya. In: The International Conference on Economics and Administration. Faculty of Administration and Business, University of Bucharest, Romania: ICEA–FAA Bucharest. Nov 2009

[8] Alaghbari WE, Kadir MRA, Salim A. The significant factors causing delay of building construction projects in Malaysia. Engineering, Construction and Architectural Management. 2007;**14**(2):192-206

[9] Hamzah N, Khoiry MA, Arshad I, Tawil NM, Ani AC. Cause of construction delay-theoretical framework. Procedia Engineering. 2011;**20**:490-495

[10] Sambasivan M, Soon YW. Causes and effects of delays in Malaysian construction industry. International Journal of Project Management. 2007;**25**(5):517-526

[11] Le-Hoai L, Lee YD, Lee JY. Delay and cost overruns in Vietnam large construction projects: A comparison with other selected countries. KSCE Journal of Civil Engineering. 2008;**12**:367-377

[12] Assaf SA, Al-Hejji S. Causes of delay in large construction projects. International Journal of Project Management. 2006;**24**(4):349-357

[13] Abd El-Razek ME, Bassioni HA, Mobarak AM. Causes of delay in building construction projects in Egypt. Journal of Construction Engineering and Management. 2008;**134**(11):831-841

[14] Al-Momani AH. Construction delay: A quantitative analysis. International Journal of Project Management. 2000;**18**(1):51-59

[15] Rahsid Y, Haq S, Aslam M. Causes of delay in construction projects of Punjab-Pakistan: An empirical study. Journal of Basic and Applied Scientific Research. 2013;**3**(10):87-96

[16] Szymański P. Risk management in construction projects. Procedia Engineering. 2017;**208**:174-182

[17] Kikwasi G. Causes and effects of delays and disruptions in construction projects in Tanzania. Australasian Journal of Construction Economics and Building-Conference Series. 2012;**1**(2):52-59

[18] Mossalam A. Projects' issue management. HBRC Journal. 2018;**14**(3):400-407

[19] Samarah A, Bekr GA. Causes and effects of delay in public construction projects in Jordan. American Journal of Engineering Research. 2016;**5**(5):87-94

[20] Al-Bahar JF, Crandall KC. Systematic risk management approach for construction projects. Journal of Construction Engineering and Management. 1990;**116**(3):533-546

[21] Zou PX, Zhang G, Wang JY. Identifying key risks in construction projects: Life cycle and stakeholder perspectives. In: Pacific Rim Real Estate Society Conference. 2006

[22] Zhu G, Bard JF, Yu G. Disruption management for resource-constrained project scheduling. Journal of the Operational Research Society. 2005;**56**(4):365-381

[23] Chileshel N, Berko PD. Causes of project cost overruns within. In: Proceedings 51 Built Environment Conference. Vol. 18. 2010. p. 20

Chapter 8

Client Type and Communication Practice during Pre-Contract Phase of Construction Project Development in South-West, Nigeria

Adesina Emmanuel Aladeloba and Godwin Iroroakpo Idoro

Abstract

This study investigates the effects of client type on communication practice adopted at pre-contract phase of construction project development. The objectives are to evaluate the levels and differences in adoption of communication practices between construction projects developed by public and private clients in South-west, Nigeria. To achieve the objectives, a sample of 394 construction projects was carried out. Three encoding-decoding communication practices: iconic, non-iconic and computer; five communication media: verbal, non-verbal, written, audio-visual and electronic and three communication channels: print, electronics and face-to-face were identified from literature. The components of the eleven communication practices were identified and measured using 5 scales. Data were collected using structured questionnaires and analyzed using percentage, mean and t-test. Results reveal non-iconic and computer as most preferred encoding-decoding communication practices by consultants engaged by public and private clients respectively and audio-visual and print as most preferred communication media and channel respectively. The study recommends further study on the effects of communication practices on project performance and an awareness of these effects among consultants. The findings of the study make significant contribution to knowledge in the area of research and adoption of performance-oriented communication practices.

Keywords: communication practice, construction projects, pre-contract phase, project client, Nigeria

1. Introduction

Communication remains an important instrument for generating project information which is regarded as a resource for determining project success while Pre-contract phase of project development is described as a documentation stage where communication practice is most required. Investigating the effects of client type on communication practice adopted at pre-contract phase of construction project development therefore, helps to place client type rightly in project performance. In effect, the levels and differences in adoption of communication practices between construction projects developed by public and private clients in South-west, Nigeria are established.

Meanwhile, the construction project contributes significantly to the development of any nation [1]. This contribution includes the workforce employment which is about 20% of the total workforce of the industry [2]. Hence it is reasonable to postulate that the industry contributes significantly to the growth of the economy through the provision of infrastructure affirmed to be a backbone for the economy. However, the inability of the construction industry to cope with the increasing complexity of construction projects informs its propensity for exceeding deadlines set for project duration [3].

These problems are linked to the nature of construction projects known to inhibit development of good communication links. Emmitt and Gorse [4] argued that timely project completion relies on the accuracy and timeliness of information exchange among the project team. As such, an appropriate information flow management system would be required to address the high degree of task uncertainties which characterize construction projects execution.

Although studies have investigated several aspects of communication in construction, the ones on "Client Type and Communication Practice at Pre-Contract Phase of Construction Project Development in South-West, Nigeria" appear to be relatively scarce. Hence, the motivation for the appraisal of levels and differences in adoption of communication practices between construction projects developed by public and private Clients during pre-contract phase of construction project delivery, with a view to choosing the most appropriate one for project delivery.

2. Literature survey

The study reviews related literature in the following areas: Nigerian construction industry, Information sharing among the project participants, Poor Communication between Client and Contractor, Communication in Construction Projects, Practice of communication in construction and Effectiveness of communication within construction teams.

Olaniran [5] asserted that the Nigerian construction industry contributes to the country's economy in numerous ways. Among them is the provision of employment for the country labour force [6, 7]. According to [8] the Nigeria construction industry remains the largest employer of labour considering the fact that over thirty percent (30%) of the country's workforce is employed by it. Samuel [9] on his part, posited that, above 69,749 workers were somehow employed by103 construction firms in the country, in the year 2018 alone.

This employment is usually by direct engagement of construction professionals and or by indirectly engaging Suppliers while purchasing construction materials [10].

The global construction industry has a value of 4 trillion dollars while the Nigeria construction industry makes up to 0.2% of the global total. Aside employment provision, [9] indicated that in comparison to other developed countries the Nigeria construction industry is relatively small.

In terms of size, [11] and [12] assert that rise in growth of the Nigeria construction industry has made it so large that same can be considered the largest in West African countries. Anny et al. [13] however opined that the growth recorded in the Nigeria construction industry is informed by the Government's continued investment in infrastructural development.

The positive impact of these sustainable practices on construction projects in developed countries, [14] elicited the sustained interest of construction professionals and academicians within the Nigerian construction industry [1].

Unfortunately, sustainable construction practice is yet to be firmly rooted in developing countries; including the Nigeria construction industry [7, 15]. Ogunde et al. [2] and Nwokoro and Onukwube [16] affirmed that sustainable practice is still a developing concept within the Nigeria construction industry. The idea behind sustainable construction emerged around 2006 following the development of the national building code in 2006 [17]. The national building code in 2006 provided an advantage for creating awareness regarding sustainable construction practice within the Nigeria construction industry, [18].

Despite the creation of the national building code, the construction industry in Nigeria still experience a lot of fragmentation [19, 20] noted that fragmentation in the Nigeria construction industry is caused due to the increasing demand for specialization.

Two major types of fragmentation namely; internal and external are rife in the Nigeria construction industry. The former which is the problem of integrating construction project team members and the latter which is attributed to the problem of integrating external bodies such as public authorities to the construction project.

Olatunji [21] however asserted that regardless the nature of fragmentation, same is responsible for the poor information sharing among project team members in the Nigeria construction industry. Abubakar et al. [22] similarly revealed that the separation of design and construction during the project development phases is informed by the fragmentation experienced within the industry. Same is responsible for project delays, poor performance, low productivity, cost overruns, disputes and many others, [23–25]. This problem, therefore, creates a paradigm shift in ensuring better communication process of construction projects' life cycle.

Yang et al., [26], asserted that building, project management requires effective collaboration and coordination between all the stakeholders in order to realize the project objectives; including successful delivery of project.

It is however pertinent to note that in procurement; whether or not it's outright project implementation, information flow continues to play a profound role in the construction environment [27]. Anything to the contrary, Wikforss and Alexander [28] declared, will affect the performance of the construction industry.

3. Communication in construction projects

According to [29], so many factors such as lack of trust and inadequate responsibility, which causes misunderstanding among two parties during construction process, affect communication between client and contractor.

Tazelaar and Snijders [30] however identified harsh relationship between client and contractor which often leads to conflicts and litigation, as major cause of communication barrier between them. Ning et al. [31] on his part, suggests that lack of public clients' initiative in relational transactions negatively influence the communication between public clients and contractors during project. Mitkus [29] identified lack of trust between client and contractor; as always resulting to argument and conflicts.

Lau and Rowlinson [32] equally mentioned lack of trust and misunderstanding as being responsible for difficulty in managing construction projects.

According to [29], the true cause of construction related conflicts is unsuccessful communication between the participants (client and contractor) in a construction project.

The absence of communication between client and contractor equally creates fears of exploitation and betrayal, which invariably results in avoidance of commitment by the team [33].

Laufer et al. [34] also mention that, poor communication between the client and contractor results to conflict, misunderstanding, uncertainty and lack of mutual cooperation among the two parties.

A study conducted by [35] highlighted that, poor communication between the parties (client and contractor) leads to estrangement and misunderstanding regarding the contract requirement while according to [36], same is one the factors that leads to time overrun in construction projects in Malaysian, Nigerian and Indian.

The construction sector is an economic investment, and its relationship with economic development is well posited. Many studies [22, 24] have discussed the contributions of the construction sector towards the economic development of nations.

The construction business is a complex enterprise with several stakeholders whose interaction requires that information be analyzed and transmitted. As such, communication remains a strategic consideration, more so that most construction disputes are due to breach or inadequate communication among the team members.

For instance, poor communication of design information often leads to design problems and eventually lead to delays and poor quality. Therefore, communication will only become effective if the receiver understands the information as intended by the sender.

At every stage of the construction lifecycle, information (in the form of drawings, specifications, notes, letters, memos, models, catalogs, instruction manuals, and pictures) needs to be stored, retrieved, and communicated.

The structure of the Dutch construction industry does not differ so much from the UK, except for the fact that Dutch industry is highly regulated in comparison with the UK [37], and that the UK subcontracting system (as opposed to the Dutch) allows for principals to contract directly to subcontractors [38].

Communication between clients and contractors during the construction project has also become a great challenge to the extent that it constitutes one of the major factors that affects project delivery. It is an important element for every organization to succeed thus making timely and accurate Communication among project stakeholders a sine-quanon in realizing the project objectives - management cannot receive information inputs, supervisors cannot give instructions, work cannot be coordinated and eventual collapse of organization [27].

Communication is a term derived from Latin word 'communis', which means to inform. It is the transformation of information; and a key function of management in any organization - ideas, goals, plans, instructions etc. are communicated to the managerial staff for the purpose of coordination. Davis [39] asserts; it is essentially a bridge of meaning between people. The communication process is made up of four key components namely: encoding, medium, decoding, and feedback while the other two factors, representing the human elements in the form of the sender and the receiver, are present.

There are multiple factors that influence the user's evaluation and perceived impacts of communication process. Trevino and Webster [40] identify flow as an important construct that characterizes perceptions of employee interactions with computer-mediated communication technologies as more or less playful and exploratory. They further hypothesized that flow is influenced by the technology (higher for electronic mail), ease of use and computer skill. Furthermore, [41] argue that communication flows vertically and horizontally in the hierarchy or it is free-flowing [42], with all the members of the organization communicating with each other.

Bartels et al. [41] further argue that communication flows vertically and horizontally in the hierarchy of organizations. It thus gives the employees the opportunity to speak out and provide critical feedback that could be important in decision-making process [43].

Communication is therefore seen as complex creative processes that occur between two or more people with the aim of exchanging information in order to solve problems and 'break new grounds'. It is however imperative that an open communication climate characterized by unrestricted dialog, honest and mutual interaction that encourages shared understanding and promotes tolerance is created, to achieve a better work environment.

4. Materials and methods

The appraisal of the Client Type is a significant factor towards ensuring the effective movements of construction information. The methodology adopted in this study to appraise difference in the levels of adoption of communication process between Public and Private Clients by construction professionals entails two primary steps namely: review of literature relating to communication process among construction professionals in the construction industry and designing questionnaire based on the reviewed literature and self-administered to the construction professionals within South-West Nigeria. The construction professionals comprise of the architect, builders, civil engineers, structural engineers, quantity surveyors, mechanical engineers and Electrical Engineers. The questionnaire was divided into two (2) sections covering the communication process and the respondent personal information. A total of four hundred and seven questionnaires (407) was received, used three hundred and ninety-four 394 (81.7%) for the analysis after scrutinizing for errors. A total of four hundred and eighty-two (482) was distributed to the respondents. Thereafter, the response to the questionnaire was inputted into Statistical Package for Social Science (SPSS) version 23. The study adopted a quantitative research design study, using Independent Sample t-test Mann-Whitney U test of difference in the levels of adoption of communication process between Public and Private Clients.

5. Analyses and discussion

407 construction projects which constituted a category of the study population, was adopted for this study. Information on the project client were investigated. The distribution of projects sampled into the sub-variables of the project client characteristic was analyzed using percentage. The results are presented in **Table 1**.

On project client, **Table 1** shows that the projects sampled are distributed among Public (22.1%) and Private Individual (77.9%). These results show that the study covered short, average and long term projects.

Characteristic	N	%
Project client		
Public	87	22.1
Private (Individual)	307	77.9
Total	394	100

N, Number of projects.

Table 1.
Descriptive results of the characteristics of project sampled.

5.1 Levels of adoption of communication process based on project client

The third research hypothesis was tested for difference in the levels of adoption of communication process between public and private clients. The hypothesis was tested using Independent Sample t-test and Mann Whitney U test at p-value ≤ 0.05. The results reveal that the differences in the levels of adoption of nine (9) communication practices between projects procured by public and private clients are significant. The results are presented in **Table 2**.

Table 2 shows that the p-values for the test of difference in the levels of adoption of written (0.032), audio-visual (0.023) and electronic communication media (0.001), print (0.001) and electronic (0.001) communication channels and upward (0.021) downward (0.001) lateral (0.007) and horizontal (0.021) communication flows between projects procured by public and private sector client are less than the critical p-value (0.05), therefore, the test rejects the hypothesis. The results indicate that the differences in the adoption of written, audio-visual and electronic communication media, print and electronic communication channels and upward, downward, lateral and horizontal communication flows between public and private sector project are significant. The implication is that project clients have significant effect on the levels of adoption of the two communication media, two communication channels and four communication flows.

6. Conclusions and recommendations

Levels of adoption of communication process between public and private sector project are significant. The implication is that project clients have significant effect on the levels of adoption of two communication media, two communication channels and four communication flows. The mean scores of the levels of adoption of the nine communication practices show that written and electronic communication media, print and electronic communication channels are more adopted in projects procured by public clients than in projects procured by private clients while audio-visual communication medium and the four communication flows are more adopted in projects procured by private clients than projects procured by public client. However, the p-values for the test of differences in the levels of adoption of iconic, non-iconic and computer encoding-decoding communication practices, verbal and non-verbal communication media, face-to-face communication channel and nexus communication flow (0.569) between projects procured by public and private sector clients is higher than the critical p-value (0.05), therefore, the test fails to reject the hypothesis.

DOI: http://dx.doi.org/10.5772/intechopen.113210

Variables tested	N	M	MR	SR	MRU	WW	Z	t-value	Df	p-value	Dec
Media:											
Written											
Public	87	3.54						2.157	392	0.032	R
Private	307	3.33									
Audio-visual											
Public	87	3.55						2.282	391	0.023	R
Private	306	3.77									
Electronic											
Public	85	2.61						3.242	389	0.001	R
Private	306	2.29									
Channel:											
Print											
Public	84	3.46						3.365	386	0.001	
Private	304	3.05									
Electronics											
Public	84	2.78						4.377	386	0.001	R
Private	304	2.35									
Flow:											
Upward											
Public	83	3.29	169.27	14049	10563	14049	-2.316			0.021	R
Private	302	3.63	199.52	60256							
Downward											
Public	83	3.22	158.45	13151	9665.5	13151	3.331			0.001	R

Variables tested	N	M	MR	SR	MRU	WW	Z	t-value	Df	p-value	Dec
Private	302	3.68	202.50	81153							
Lateral											
Public	83	3.35	165.24	13715	10229	13715	2.681			0.007	R
Private	302	3.77	200.63	60590							
Horizontal											
Public	85	3.15	161.10	13608	9953.5	13608	3.277			0.001	R
Private	301	3.60	202.93	61082							
Nexus											
Public	85	3.35	187.64	15949.5	122945.5	15949	0.570			0.569	FR
Private	301	3.45	195.15	58741.5							

N, Number of respondents; M, Mean; MS, Mean score; MR, Mean rank; SR, Sum of ranks; MWU, Mann–Whitney U; WW, Wilkinxon W; Df, Degree of freedom; Dec, Decision; R, Reject; FR, Fail to reject.

Table 2.
Results of independent sample t-test Mann-Whitney U test of difference in the levels of adoption of communication process between public and private clients.

The results indicate that the difference in the levels of adoption of the communication processes between public and private sector projects is not significant; therefore, the type of client does not have significant effect on the levels of adoption of the communication processes between public and private sector projects during pre-contract phase of project delivery.

Clients and consultants should discourage the adoption of computer encoding-decoding practice, written communication medium and nexus communication flow during pre-contract phase of project delivery in the efforts to minimize delay in the delivery of projects.

Clients and consultants should give high consideration to communication speed, quality of project information, expected frequency of communication of project information, number of project participants, technology to be used for communicating project information in their choice of appropriate communication process to be adopted during pre-contract phase of project delivery.

Project clients and consultants should give consideration to the adoption of iconic and non-iconic encoding-decoding practices when procuring projects by design-build method, the adoption of upward, downward and lateral practices when procuring projects by design-bid-build method and the adoption of electronic communication medium and print, face-to-face and electronics communication channels when procuring projects by labour-only method.

Acknowledgements

The authors like to appreciate everybody that has contributed to this article directly or indirectly.

Conflicts of interest

The author hereby declares that there is no conflict of interest.

Author details

Adesina Emmanuel Aladeloba[1*] and Godwin Iroroakpo Idoro[2]

1 Department of Building Technology, Yaba College of Technology, Lagos, Nigeria

2 Department of Building, University of Lagos, Lagos, Nigeria

*Address all correspondence to: aladelobasina@yahoo.com

References

[1] Daniel EI, Oshineye O, Oshodi O. Barriers to sustainable construction practice in Nigeria. In: Proceedings of the 34th annual ARCOM Conference 3-5 September 2018. Belfast, UK: ARCOM; 2018. pp. 149-158. Available from: https://www.researchgate.net/publication/373444857

[2] Ogunde A, Olaolu O, Afolabi A, Owolabi J, Ojelabi R. Challenges confronting construction project management system for sustainable construction in developing countries: Professionals perspectives (a case study of Nigeria). Journal of Building Performance. 2017;**8**:1-11

[3] Tukel O, Rom W, Kremic T. Knowledge transfer among projects using a learn-forget model. The Learning Organization. 2007;**15**:179-194

[4] Emmitt S, Gorse C. Construction Communication. Oxford, UK: Blackwell Publishing Ltd.; 2003

[5] Olaniran H. On the role of communication in construction projects in Nigeria. Journal of Environmental Science and Technology. 2015;**2**:048-054

[6] Ibrahim II, Githae W, Stephen DB. Indigenous contractors' involvement and performance in construction procurement systems in Nigeria. Global Journal of Research in Engineering. 2014;**14**. Avilable from: https://api.semanticscholar.org/CorpusID:33814090

[7] Dahiru D, Abdulazeez A, Abubakar M. An evaluation of the adequacy of the national building code for achieving a sustainable built environment in Nigeria. Research Journal of Environmental and Earth Sciences. 2012;**4**:857-865

[8] Iyagba R, Mafimidiwo B. Comparative study of problems facing small building contractors in Nigeria and South Africa. Journal of Emerging Trends in Economics and Management Sciences. 2016;7:101-109

[9] Samuel B. The Nigerian Construction Industry outlook report of 2018. 2018. Available: https://brandspurng.com/2017/10/12/the-nigerian-construction-industry-outlook-2018/ [Accessed: Feburary 2, 2018]

[10] Fapohunda TM. Women and the informal sector in Nigeria: Implications for development. British Journal of Arts and Social Sciences. 2012;**4**:35-45

[11] Balogun S, Esezobor D, Adeosun S, Sekunowo O. Challenges of producing quality construction steel bars in West Africa: Case study of Nigeria steel industry. Journal of Minerals and Materials Characterization and Engineering. 2009;**8**(4):283-292. DOI: 10.4236/jmmce.2009.84025

[12] Ede A. Building collapse in Nigeria: The trend of casualties in the last decade (2000-2010). International Journal of Civil & Environmental Engineering IJCEE-IJENS IJENS I J E N S. 10 January 2010; pp. 6-32

[13] Anny AN, Anthony CI, Kehinde OM. Critical issues in reforming the Nigerian construction industry. British Journal of Applied Science & Technology. 2015;**5**:321

[14] Ofori G. Construction in developing countries. Construction Management and Economics. 2007;**25**:1-6

[15] Aghimien D, Fadeke A, Aghimien E, Awodele O. Challenges of sustainable construction: A study of educational

buildings in Nigeria. Journal of Performance of Constructed Facilities. 2018;**14**:71-74

[16] Nwokoro I, Onukwube HN. Sustainable or green construction in Lagos, Nigeria: Principles, attributes and framework. Journal of Sustainable Development. 2011;**4**:166

[17] Ogunbiyi M. The national building code and the construction industry professionals in Nigeria. International Journal of Social Sciences and Entrepreneurship. 2014;**1**:937-948

[18] Oladokun M, Adewuyi T. Implications of the national building code to sustainable growth of indigenous construction firms in Nigeria: A qualitative research approach. In: 1st International Conference of the School of Management Technology (SMAT). Akure, Nigeria: Federal University of Technology; 28th-31st March, 2017

[19] Obiegbu M. Understanding the National Building Code–Utilizing the provisions of the code in practice of building profession. In: A Paper Delivered At 2007 Annual General Meeting of Abia State Chapter of Nigerian Institute of Building (NIOB). 2007

[20] Mohd Nawi MN, Baluch N, Bahaudin AY. Impact of fragmentation issue in construction industry: An overview. In: MATEC Web of Conferences. Building Surveying, Facilities Management and Engineering Conference (BSFMEC 2014). 2014. DOI: 10.1051/matecconf/20141501009

[21] Igbaekemen GO. The impact of information communication technology on small and medium scale enterprises productivity in Nigeria. European Journal of Computer Science and Information Technology. UK: ECRTD. April 2020;**8**(2):23-37. ISSN: 2054-0957 (Print), Online ISSN: 2054-0965 (Online)

[22] Abubakar J, Ibrahim M, Kado YD, Bala K. Contractors' perception of the factors affecting Building Information Modelling (BIM) adoption in the Nigerian Construction Industry. International Conference on Computing in Civil and Building Engineering. 2014;**2014**. DOI: 10.1061/9780784413616.022

[23] Kehinde JO, Mosaku TO. An empirical study of assets structure of building construction contractors in Nigeria. Engineering, Construction and Architectural Management. 2006;**13**:634-644

[24] Adamu M, Olufemi Bioku J, Basiru Kolawole O. Assessing the characteristics of Nigerian construction industry in infrastructure development. International Journal of Engineering Research & Technology (IJERT). Nov 2015;**04**(11)

[25] Nnadi E, Enebe E, UGWU O. Evaluating the Awareness Level of Risk Management amongst Construction Stakeholders in Nigeria. International Journal of Construction Engineering and Management. 2018;**7**:47-52

[26] Yang J, Ahuja V, Shankar R. Managing Building Projects through Enhanced Communication – An ICT Based Strategy for Small and Medium Enterprises, CIB World Building Congress 2007. South Africa: CIB; 2007. pp. 2334-2356

[27] Titus S, Bröchner J. Managing information flow in construction supply chains. Journal of Construction Innovation: Information, process, management. 2005;**5**:71-82. Available from: https://api.semanticscholar.org/CorpusID:108877513